A History of Accident and Emergency
Medicine, 1948–2004

Also by Henry Guly:

DIAGNOSTIC ERRORS IN TRAUMA CARE

HISTORY TAKING, CLINICAL EXAMINATION AND RECORD KEEPING IN EMERGENCY MEDICINE

A History of Accident and Emergency Medicine, 1948–2004

Henry Guly
Consultant in Accident and Emergency Medicine
Derriford Hospital
Plymouth

© Henry Guly 2005

All rights reserved. No reproduction, copy or transmission of this publication may be made without written permission.

No paragraph of this publication may be reproduced, copied or transmitted save with written permission or in accordance with the provisions of the Copyright, Designs and Patents Act 1988, or under the terms of any licence permitting limited copying issued by the Copyright Licensing Agency, 90 Tottenham Court Road, London W1T 4LP.

Any person who does any unauthorised act in relation to this publication may be liable to criminal prosecution and civil claims for damages.

The author has asserted his right to be identified as the author of this work in accordance with the Copyright, Designs and Patents Act 1988.

First published in 2005 by
PALGRAVE MACMILLAN
Houndmills, Basingstoke, Hampshire RG21 6XS and
175 Fifth Avenue, New York, N.Y. 10010
Companies and representatives throughout the world.

PALGRAVE MACMILLAN is the global academic imprint of the Palgrave Macmillan division of St. Martin's Press, LLC and of Palgrave Macmillan Ltd. Macmillan® is a registered trademark in the United States, United Kingdom and other countries. Palgrave is a registered trademark in the European Union and other countries.

ISBN-13: 978–1–4039–4715–4
ISBN-10: 1–4039–4715–5 (hardback)

This book is printed on paper suitable for recycling and made from fully managed and sustained forest sources.

A catalogue record for this book is available from the British Library.

Library of Congress Cataloging-in-Publication Data

Guly, H. R. (Henry R.)
   A history of accident and emergency medicine, 1948–2004 / by Henry Guly.
   p. cm.
   Includes bibliographical references and index.
   ISBN 1–4039–4715–5 (hardback)
   1. Emergency medicine – Great Britain – History – 20th century.
2. Hospitals – Great Britain – Emergency service – History – 20th century.
I. Title.
   [DNLM: 1. Emergency Medicine – history – Great Britain. 2. Emergency Service, Hospital – history – Great Britain. 3. History of Medicine, 20th Cent. – Great Britain. 4. History of Medicine, 21st Cent. – Great Britain. WX 215 G973h 2005]
RC86.7.G855 2005
616.02'5—dc22
                                                      2004051499

10 9 8 7 6 5 4 3 2 1
14 13 12 11 10 09 08 07 06 05

Printed and bound in Great Britain by
Antony Rowe Ltd, Chippenham and Eastbourne.

This book is dedicated to the pioneers of casualty/accident and emergency medicine/emergency medicine

Contents

List of Tables	ix
List of Figures	x
Acknowledgements	xi
Preface	xii
List of Abbreviations	xiv
Some Notes on the Statistics and Other Topics	xvi
1 Casualty Staffing before Platt	1
2 Who Should Run A&E Departments?	27
3 The First Consultants	50
4 Senior Registrars and Training	64
5 How Many Consultants?	74
6 A Changing Specialty	87
7 Academic A&E, the Faculty and Changes of Name	100
8 Non-consultant and Non-training-grade Doctors	111
9 Junior Staffing of A&E Departments	122
10 Primary Care in A&E	135
11 Politics and the Future	151
Appendix A: Members Present at the First Committee Meeting of the Casualty Surgeons Association 12 October 1967	156
Appendix B: Officers of Casualty Surgeons Association and British Association for Accident and Emergency Medicine	157
Appendix C: The First Board of the Faculty of Accident and Emergency Medicine 1993	158

*Appendix D: Officers of the Faculty of Accident and
Emergency Medicine* 159

References 160

Index 179

List of Tables

1.1	Number of casualty departments in 1960 by hospital size (Platt Report)	6
1.2	Size of casualty departments in 1960	6
1.3	Appointments of SCOs	13
1.4	Preferred career specialty for SCOs 1962	14
1.5	Staffing levels in A&E departments	16
1.6	Grade of most senior casualty post in departments without an SCO 1958	19
1.7	Rota for out of hours cover included	19
1.8	Most senior doctor seeing new patients	20
1.9	Who was in charge? – Two surveys 1958	22
2.1	Supervision of casualty departments 1969	38
2.2	Staffing of 228 A&E departments 1969	39
3.1	Number of A&E consultants	53
4.1	A&E senior registrar numbers	66
4.2	Higher qualifications of the senior registrars in post February 1981	68
4.3	Problems encountered by SRs after becoming consultant (affecting over 50 per cent)	71
4.4	The work of an A&E registrar in a year	71
5.1	Projected staffing of A&E departments in Devon in 1981	76
5.2	BAEM consultant staffing recommendations 1993	84
6.1	Workload of A&E consultants compared to all consultants 1990	93
6.2	Hours worked by nine A&E consultants 1998	94
8.1	Clinical assistants in A&E 1984	116
8.2	Staff grade appointments in A&E 1996	119
9.1	Percentage of medical school graduates choosing general practice as a career	128
9.2	A&E SHO country of origin in 1995	131
9.3	Percentage of A&E SHOs on rotational posts in 1995	132
9.4	Career intentions of A&E SHOs in 1995	132
9.5	Staffing of A&E departments 1999	133

List of Figures

1.1	Quality of the casualty officer	19
1.2	Quality of out-of-hours cover	20
1.3	Standard of consultant cover	24
1.4	Interest of Medical Committee	25
4.1	Number of registrars, senior registrars and specialist registrars in A&E in England 1973–2002	73
5.1	Number of consultants in A&E in England 1973–2002	83
5.2	A&E consultants in England as percentage of total consultants 1973–2002	84
8.1	Number of associate specialists, clinical assistants, hospital practitioners and staff grades working in A&E in England 1973–2001	112
9.1	Number of HOs in A&E in England 1973–2001	124
9.2	Number of SHOs in A&E in England 1973–2001	125

Acknowledgements

My main acknowledgement must be to the Wellcome Trust who awarded me a Research Fellowship which allowed me to spend four months in researching and writing this book. I must also acknowledge the assistance of Dr Mark Jackson of the Centre for Medical History, Exeter University for his guidance during the research. I also need to thank:

Roger Evans, Past President of BAEM for allowing me to use and quote from the BAEM minutes and other papers.
Julie Bloomfield, administrator of the BAEM.
Mandy Mordue Archivist BMA House for allowing me access to the minutes of the SCO Subcommittee, the CCSC, the CCHMS and the JCC.
The librarians at Derriford Hospital, Plymouth for assisting me in obtaining obscure reports, papers etc.

In the research for this book I have spoken to, interviewed, written to or emailed many individuals and would like to thank the following for their help:
Chief Executive, British Orthopaedic Association, Edward Abson, Howard Baderman, David Cain, Peter Cox, Christopher Cutting, Andrew Dove, David Fergusson, Judith Fisher, Kambiz Hashemi, Miles Irving, Keith Little, Stuart Lord, Jonathan Marrow, Andrew Marsden, Alasdair Matheson, Stephen Miles, Patrick Nee, Mark Prescott, Tony Redmond, Iain Robertson-Steel, William Rutherford, Ian Stewart, Ian Swann, William Tullett, David Wilson, Frederick Wright and David Yates.

Preface

Accident and emergency (A&E) medicine started as 'casualty' and will almost certainly soon be called 'emergency medicine', the name by which the specialty is known in the USA and Australasia. A&E is a curious specialty (and for many years it was argued that it was not a specialty) in that, while almost every other specialty originates in increased subspecialisation, A&E has remained very general, covering the acute care of all emergencies, including injuries, in patients of all ages and suffering from diseases of all body systems. Doctors who form a new specialty are usually already consultants but A&E was largely developed by doctors in a subconsultant grade. Most specialties, while they overlap other specialties, are based on a core of knowledge or skill which is specific to that specialty. A&E is different as A&E doctors do nothing which cannot be done by others. Thus A&E doctors repair wounds but this is also undertaken by surgeons of various types; cardiac resuscitation is also performed by physicians. What is unique about A&E is the broadness of the specialty and the organisation and triage which allows that to happen.

That there needs to be a specialty of A&E is shown by the fact that emergency medicine in the USA and Australia developed at almost exactly the same time as A&E developed in the UK. In the USA the American College of Emergency Physicians was formed in 1968[1] and in Australia the first full time consultant in charge of an emergency department was appointed in 1967 and the Australian Society of Emergency Medicine was formed in 1981 with a College forming in 1984.[2]

Although A&E remains a broad specialty, its emphasis has changed considerably in the past 30 years. What started as a surgically orientated specialty heavily involved with the trauma which forms the bulk of the workload, is moving towards medicine, concentrating on resuscitation and the medical emergencies which form the majority of the seriously ill patients who attend A&E. While a number of the early consultants did some orthopaedics (including operative orthopaedics), many younger consultants have a background in medical specialties.

My introduction to the specialty (apart from a brief contact as a patient during childhood) was as a medical student in 1971. I was an SHO in 1976 and was one of the early senior registrars in 1980. I have been a consultant since 1983 in two hospitals in different parts of the country. I have also played a small role in the developing specialty, having served at different times on the Casualty Surgeons Association and British Association for Accident and Emergency Medicine executive committees and the Board of the Faculty of Accident and Emergency Medicine. I have therefore observed the development of the specialty at close hand.

The full history of A&E has yet to be written. To do it justice would require a large volume to describe facilities, A&E nursing, workload, changing case-mix, major accidents, research, flying squads and the improvements in care which have occurred. Improvements include things as varied as the description of the Glasgow Coma Scale, CT scanning and thrombolysis.

In an editorial detailing some of the problems in A&E departments in 1969, the *Lancet* commented: '... when the emergency services are in the hands of experienced and interested doctors, the problems discussed ... will take care of themselves.'[3] The following year it said: 'The crucial first step in reform must be to ensure that every department has at its head a consultant whose duties lie primarily or entirely within it.'[4] I have therefore tried to describe probably the most important part of the story of A&E. This is the battle to get the specialty recognised with consultants at the head of every department and I describe the medical staffing of departments in general. To do this has required going off at numerous tangents. For example one cannot discuss consultant numbers without describing what consultants do and one cannot discuss GPs in A&E without discussion of the primary care problems which present.

Describing medical politics and facts is relatively easy as one can report that a meeting occurred on a specific date or detail how many consultants were in post in a particular year. Describing medical advances is more difficult. For example, some A&E departments (within the same hospital as a neurosurgical unit) had use of CT scanning by the late 1970s whereas others did not have this facility for another 15 years. Most difficult of all is to describe attitudes to, and within, the emerging specialty. For this I have made use of quotes from letters to journals and have also based it on my own experience.

I thank all those who gave me help in this history. The responsibility for any errors and for the interpretation put on some of these facts, is mine alone.

List of Abbreviations

A&E	Accident and emergency
AAC	Appointments advisory committee
ACLS	Advanced Cardiac Life Support
APLS	Advanced Paediatric Life Support
AS	Associate specialist
ATLS	Advanced Trauma Life Support
BAEM	British Association for Accident and Emergency Medicine (from 2004, British Association for Emergency Medicine)
BMA	British Medical Association
BMJ	British Medical Journal
BOA	British Orthopaedic Association
CA	Clinical assistant
CAS	Clinical assessment system
CCHMS	Central Committee for Hospital Medical Services (of the BMA)
CCSC	Central Consultants and Specialists Committee (of the BMA)
CCST	Certificate of Completion of Specialist Training
CSA	Casualty Surgeons Association
DGH	District general hospital
DHSS	Department of Health and Social Security
DoH	Department of Health
EMJ	Emergency Medicine Journal
EMRS	Emergency Medicine Research Society
FAEM	Faculty of Accident and Emergency Medicine
FFAEM	Fellow of the Faculty of Accident and Emergency Medicine
FRCP	Fellow of the Royal College of Physicians
FRCS	Fellow of the Royal College of Surgeons
FRCS(Ed)	Fellow of the Royal College of Surgeons of Edinburgh
GP	General practitioner
HO	House officer
HS	House surgeon
ICU	Intensive care unit
JCC	Joint Consultants Committee
JCHMT	Joint Committee for Higher Medical Training
JCHST	Joint Committee for Higher Surgical Training

JCHTA&E	Joint Committee for Higher Training in Accident and Emergency
JCHTAnaes	Joint Committee for Higher Training in Anaesthetics
JHMO	Junior hospital medical officer
MA	Medical assistant
MFAEM	Member of the Faculty of Accident and Emergency Medicine
MIU	Minor injuries unit
MRCGP	Member of the Royal College of General Practitioners
MRCP	Member of the Royal College of Physicians
MRCS	Member of the Royal College of Surgeons
MTOS	Major Trauma Outcome Study
NCCG	Non-consultant career grade (doctor). A term used to describe associate specialists, staff grade doctors and similar doctors employed on Trust contracts
PRHO	Pre-registration house officer
RCA	Royal College of Anaesthetists
RCP	Royal College of Physicians
RCS	Royal College of Surgeons
RMO	Resident medical officer
RSO	Resident surgical officer
SAC	Specialist Advisory Committee
SCO	Senior casualty officer
SAS	Staff and associate specialists
SCOPME	Standing Committee on Postgraduate Medical Education
SMHO	Senior hospital medical officer
SHO	Senior house officer
SG	Staff grade
SpR	Specialist registrar
SR	Senior registrar
STA	Specialist Training Authority
TARN	Trauma Audit and Research Network
UCH	University College Hospital
WTE	Whole time equivalent

Some Notes on the Statistics and Other Topics

Statistical information is essential to understand the growth of a specialty but it is not easy to obtain. NHS figures for England, England and Wales, Scotland and Northern Ireland are all held in different places. These sources ignore doctors working in the armed forces who are outside the NHS. There is also no doubt that some of the information is actually wrong and this is discussed in individual chapters.

There are several sources of statistics on medical staffing in hospitals.

My graphs use information on medical staffing in England alone. This was obtained from the internet at www.doh.gov.uk/stats/history.htm with figures for 1973 and 1974 being obtained directly from the Department of Health. Before that date, figures for A&E were combined with those for trauma and orthopaedics. These statistics have the advantage that they include numbers of pre-registration house officers, clinical assistants and hospital practitioners. They do have the disadvantage, not only that they are just from England, but that after 1986 the numbers of registrars and senior registrars are combined and not given separately.

The Statistical Bulletin publishes yearly details of hospital, public health medicine and community services medical and dental staff in England. Unfortunately these figures are rounded off to the nearest five. As A&E is a small specialty, this can give a false impression of numbers.

Each year the Medical Manpower Division of the DHSS (later the Medical Manpower and Education Division of the DoH) used to publish details of medical staffing in England and Wales in the journal *Health Trends*. This has the advantage of covering England and Wales rather than just England. It also has the advantage of better breakdown of staffing numbers by sex and of maintaining separation of registrar and senior registrar numbers for longer. A final advantage is that for a few years they gave separate figures for consultants who practised only in one specialty and consultants who practiced in more than one (though this group is very small). The disadvantage of the numbers in *Health Trends* is that they were written to aid career planning and so do not include numbers of clinical assistants or hospital practitioners.

None of the official sources give figures for non-standard grades of doctors such as Trust Grades.

Lastly figures are often quoted in various documents and reports. These will usually be the figures for England and Wales. However some may be the result of surveys covering the whole of the UK. The other disadvantage of surveys is that few have a 100 per cent return rate.

In A&E the differences between number of post holders and whole time equivalents (WTE) is very small (see Figure 5.1 for the differences for consultants). For convenience most of the graphs show post holders only. The exception is for clinical assistants and hospital practitioners which are designed to be part-time appointments and so the graphs showing these, give WTE.

What started as 'casualty' departments officially became 'accident and emergency' departments or 'accident' departments after 1962 though the word 'casualty' remains in common use even today. In many places they are now called 'emergency' departments. The specialty itself has changed from 'casualty' to 'accident and emergency medicine' with many now calling it 'emergency medicine'. I regard the terms synonymously and in general have used the word 'casualty' to describe departments before 1962 and the specialty until the first consultants were appointed. After that I have tended to use 'A&E' There may be places where it seemed more appropriate to use another term and, obviously, when quoting others, I have used their words. I confess to making minimal changes in a few of the quotes. 'Accident and emergency' has been abbreviated 'A&E', 'A and E', 'A & E', A + E and even 'A/E'. For consistency I have always abbreviated to A&E. Similarly the use of capital letters in for example, 'Accident and Emergency Department' varies from author to author and I have tried to apply a consistent approach to capital letters.

The organisation supervising the NHS started as the Ministry of Health and then became the Department of Health and Social Security (DHSS) and, later still, the Department of Health (DoH). There is, currently, a splitting of responsibilities between the DoH and the NHS Executive. I hope I have applied the correct term at different stages of the history but while differences between these organisations may be very important to those interested in the administration of the NHS, they are unimportant in the history of A&E.

Books for further reading

Abel-Smith B. *The Hospitals 1800–1948*. (1964) Heinemann, London.
Department of Health. *Reforming Emergency Care*. (2001) Department of Health, London.

Lowden TG. *The Casualty Department*. (1955) E&S Livingstone, Edinburgh.
Nuffield Provincial Hospital Trust. *Casualty Services and their Setting*. (1960) Oxford University Press, Oxford.
Standing Medical Advisory Committee. *Accident and Emergency Services (The Platt Report)*. (1962) HMSO.

1
Casualty Staffing before Platt

The most influential report into casualty or accident and emergency (A&E) services was the Standing Medical Advisory Committee Report on Accident and Emergency Services chaired by Sir Harry Platt and published in 1962.[1,a] This was a major milestone in the history of A&E and this chapter aims to describe medical staffing of casualty departments between the start of the NHS and that report.

What is in a name?

The noun 'casualty' and related adjective 'casual' both have several meanings and this has served to confuse and inflame discussion on the history and function of casualty departments. Before describing the history, some discussion on the words is essential. The Shorter Oxford Dictionary[2] gives two definitions of the word 'casual' dating back to the period 1350–1469. These are:

1. Due to, characterised by or subject to chance; accidental, fortuitous. Non-essential.
2. Occurring unpredictably, irregular; occasional.

The Dictionary states that 'casualty', in the same period, meant 'a chance occurrence, an accident, a mishap, a disaster' but by the

---

[a] Sir Harry Platt also chaired working parties on other topics and there are several documents known as Platt Report but in this book, the words refer to the report on accident and emergency services unless qualified. In addition Sir Robert Platt also wrote a report (*Joint Working Party (Chairman Sir Robert Platt). Medical Staffing Structure in the Hospital Service. 1961, HMSO*) which features in the history of A&E medicine and which is also sometimes known as 'Platt Report'. They were known in casualty circles as the 'Sir Harry Platt Report' and the 'Sir Robert Platt Report'.

mid-nineteenth century another meaning of the word had emerged as: 'A person killed or injured in a war or an accident; a thing lost or destroyed.' The military use of the word also covered those unfit for battle because of illness.[3]

By the late nineteenth century the word 'casual' had also taken on additional meanings including: 'occurring or brought about without design or premeditation; having no specific plan, method, motivation or interest. Also (of a person, action, etc.), unmethodical, careless, unconcerned, uninterested, informal, unceremonious.'

Origins of casualty departments

While the aim of this chapter is to describe casualty departments and their staffing from the start of the NHS in 1948 until 1962, it is important that they are put into historical context. Ever since the first hospitals were built, casualties (i.e. those injured and suddenly taken ill) would have presented at hospital for treatment. Abson[4] states that '... "casualty" was introduced into hospital administration 400 years ago to designate those who, overtaken by some chance calamity, were immediately rendered "community dependent"'. He gives no reference and even if the word had been used for that length of time, it was not in common usage as at St Bartholomew's Hospital the word was first applied to patients in 1824,[5] and the Shorter Oxford Dictionary defines a casualty department or ward as 'the part of a hospital where casualties are attended to' and dates this use as the mid-nineteenth century. Whether a patient was a casualty or not had important implications as in many voluntary hospitals, accident victims[6] and casualties could be admitted directly by a physician or surgeon whereas other patients needed a letter of recommendation from one of the governors or subscribers.[7] Sorting and disposal of patients would have initially been done on a ward but as hospitals became busier it became more convenient to provide separate accommodation near the hospital entrance for these patients.[8]

Casualty departments had another origin in the arrangements made to treat casual attenders. Ell, in an article on the origins of the word 'casualty' states that the 'casual attender' originated as the 'workhouse casual, who was not one of the unemployable permanents, but the irregular and unexpected caller who needed temporary help: when the workhouse became the poor-law hospital, what more natural to label the odd sick who appeared for outpatient help without appointment, also as casuals.'[3]

Different interpretations of the words 'casualty' and 'casual' have been implied by those wishing to promote or to denigrate casualty departments. Ell says that 'it is more respectable to be a military casualty than a workhouse casual, and it is more estimable to be a doctor who deals with casualties than with casuals'.[3] Those who really want to denigrate the departments may imply that the staff or treatment is casual.

Until the mid-nineteenth century the volume of outpatient work was very small and was a token gesture rather than a means of providing health care. A history of St Bartholomew's Hospital states that in 1678 its outpatient workload had become burdensome and was limited to eight patients per week though this was increased to 50 per week in 1696 by popular demand.[9] Outpatient work in all hospitals grew considerably from about 1835–1850 and a large percentage of this was casualty work.[10] Few were accidents and most were casual attenders. This caused opposition from general practitioners (GPs) as outpatient departments, by providing free treatment, were said to rob the GP of his livelihood.[6,10]

By the nineteenth century decisions on who to admit to hospital rested with doctors rather than the governors and the differentiation between casualties and other outpatients was not well defined. In 1869 of the outpatients department at St Bartholomew's, it was said that 'the patients are divided into two categories...They are the "casualty" which comprises those who are supposed to require temporary treatment for diseases or injuries of a trifling character, and the "outpatients" properly so called who after receiving a letter of admission are entitled to the advice of the assistant surgeons and physicians...'[11]

In 1910 a report from the King Edward's Hospital Fund said: 'outpatients can be divided into several classes...one of (which) is that known as "casualties". This term has no fixed definition common to all hospitals but it usually includes cases which are either too urgent or too trivial to be referred to the outpatient department proper at the hours of attendance of the visiting staff...'[11]

This seems to summarise casualty departments in the first half of the twentieth century: they dealt with the very urgent cases (injuries and illness) and also a lot of very minor illness and by doing this, casualty departments were the main providers of care for the poor.[12] This continued to bring casualty departments and general practitioners into conflict.[6] Before the NHS many hospitals ran a casualty department but not always by that name. Clarkson in 1949 in an article about the casualty department at Guy's Hospital describes himself as a casualty surgeon but apart from that, the word 'casualty' is not used. The department is called the 'Front Surgery' and the doctors working there are described as 'outpatient officers'.[13] Other names used in different hospitals were: 'Receiving Room', 'Receiving Hall',[14] 'Sorting Room'[6] and SOPD (Surgical Out Patient department).[15] Different casualty departments also had varied relationships with other parts of the hospital. Thus some were closely linked with surgery and provided elective minor surgical procedures and others were also used as a receiving room for patients sent in by their general practitioner for admission.

1948–1962

Between 1948 and 1960 there was little of substance in the medical literature describing casualty services. There was a series of articles in the *Lancet* in 1956 by Lowden from Sunderland,[16–18] and there were other significant articles by Patrick Clarkson from Guy's Hospital which are quoted below. These two consultants seem to have been exceptional in the interest they took in their departments. For quantitative information, by far the best source of information is a large study of casualty services done by the Nuffield Hospitals Trust in 1958 and published in 1960,[11] described by the *Lancet* at the time as 'the most comprehensive examination of this part of the hospital service that has yet appeared in print'.[19] Research for this study included questionnaires to a large number of hospitals and visits to 18 departments chosen to be representative of different areas and types of department.

In the 1950s the extent of the functions of a casualty department varied from place to place.[20] Clarkson[21] said that there were many different types of casualty department but that they divided into two main types. In a comprehensive department, all the sick and injured passed through a common hospital portal (though there might be segregation within the department). He says that this occurred for example at Guy's Hospital, the Royal Northern Hospital and some other London hospitals. The other type of casualty department he describes is the sectional or segregated department where different groups of patients were separated at the entrance and were treated in different departments. The usual segregation was between the sick and the injured as occurred at, for example, the London Hospital, the Radcliffe Infirmary, Oxford and the Birmingham Accident Hospital. Other forms of segregation were between medical and surgical (including accidents). This system was in operation at the Middlesex Hospital in the mid-1950s when patients with obviously surgical conditions, such as road accidents and septic hands were seen in a separate surgical department and all the rest seen by one of two casualty medical officers.[22] I witnessed similar arrangements at St Mary's Hospital, Paddington as a medical student in 1971 where surgical patients were seen by the casualty officers with medical patients (at least during working hours), being seen by the duty medical registrar. Further segregation of patients into gynaecology and children also occurred in places (e.g. at the Edinburgh Royal Infirmary).[21] In the segregated department, patients needed to be sorted or triaged into appropriate groups: this was probably done differently at different hospitals by nursing and reception staff. Caro (later the first A&E consultant at St Bartholomew's Hospital) described a popular story (which he hoped was untrue) of triage done by a porter who had been taught that oedema of one leg is surgical and oedema of two legs is medical.[23]

Casualty departments were much smaller than modern A&E departments. Not only did fewer patients attend but also there were many more departments, so patients were more thinly spread. In Plymouth in the 1950's there were three casualty departments at Devonport, Freedom Fields and Greenbank Hospitals, the latter two being no more than 250 yards apart. Portsmouth in 1960 also had three casualty departments which were described in an article in the *British Medical Journal* (*BMJ*). One of these at the Royal Hospital was the major department seeing 44,012 patients in total but St Mary's Hospital and Queen Alexandra's Hospital both saw between five and six thousand patients. In addition there were cottage hospitals without resident staff and the Portsmouth Eye and Ear Hospital saw casualties in those specialties.[24]

In 1960, in 2644 hospitals of different sorts, the Platt Report[1] found 789 casualty departments which saw at least one patient per week. Nineteen of these were in eye hospitals and 26 in children's hospitals. Hospitals were smaller and the differentiation between a general hospital and a cottage hospital was less marked. Table 1.1 shows the number of casualty departments in relation to hospital size. Of the 789 departments, only 31 saw more than 500 new patients a week, equivalent to 26,000 new patients per year. Table 1.2 shows the size of the casualty departments.

Arrangements within the hospital were also very different from now. One hospital inspected in 1958 for the Nuffield Report had no orthopaedic surgeon within the hospital group with all fractures being dealt with by the six general surgeons. In another hospital there was no opportunity to refer a patient to a fracture clinic run by anyone. In Weston-super-Mare there was only a single orthopaedic surgeon who alternated with one of the four general surgeons to deal with all the major injuries to the limbs and the spine.[25] However it must be remembered that

Table 1.1 Number of casualty departments in 1960 by hospital size (Platt Report)

Hospital size (no. of beds)	Number of casualty departments
0–25	150
26–50	150
51–100	120
101–250	191
Over 250	178
Total	789

Source: Platt Report.

Table 1.2 Size of casualty departments in 1960

New patients per week	Number of casualty departments
Under 100	467
100–199	140
200–299	80
300–399	47
400–499	24
Over 500	31

Source: Platt Report.

general surgeons at that time would have dealt with orthopaedic trauma in their training and most would have had experience of treating wartime casualties. Of the 789 hospitals which had casualty departments, only 535 had a fracture clinic[1] and even in the 31 hospitals with casualty departments seeing over 500 new patients per week, only 29 had fracture clinics. In many cases, patients seen in casualty in one hospital could be referred to a fracture clinic in another hospital. Lowden[20] in a book on casualty published in 1955 describes working in a department where orthopaedic injuries were treated in a separate accident and fracture hospital and as a consultant general surgeon working in casualty, he concentrated on sepsis, soft tissue injuries and wounds and did not see it as his role to treat fractures.

Accident services

At the turn of the twentieth century orthopaedics in the UK was a relatively unimportant part of surgery and its field was limited to the correction of deformity[26] with injured patients looked after by general surgeons. The first fracture service was established by Robert Jones (who for many years had the title of general surgeon) for the workmen building the Manchester Ship Canal between 1888 and 1893.[27] During the First World War some military orthopaedic centres were established under the leadership of Robert Jones, where certain categories of patients were looked after by orthopaedic surgeons and this was the start of orthopaedic interest in trauma. After the war the care of injured civilians largely reverted to the care of general surgeons but some orthopaedic surgeons continued with managing trauma and set up fracture clinics. The first of these was established by Harry (later Sir Harry) Platt at Ancoats Hospital, Manchester. Patients managed in fracture clinics by orthopaedic surgeons had considerably better results than those managed by general surgeons[28] and in 1935 a Committee on Fractures recommended that the care of patients with fractures needed standardisation and a proper organisation. Their four principles of the care of fractures were: segregation of fractures from other patients; continuity of treatment from the primary reduction of the fracture through to the end of rehabilitation; proper aftercare of fractures and unity of control, that is, the care of the patient through out all the phases of treatment being the responsibility of a single person.[28]

In 1939 the Interdepartmental Committee of the Home Office, the Ministry of Health and the Scottish Office on the Rehabilitation of Persons Injured by Accidents reported: '...the system...by which fractures are treated in general surgical wards under general surgical routine, is gravely defective. ... A radical change of method is necessary'.[29] In 1943 the British Orthopaedic Association (BOA), in its first memorandum on accident services recommended that the emphasis on fractures needed to be extended to all injuries. They recommended that accident services must be developed by surgeons widely trained in orthopaedics and trauma and that 'the Director of an Accident Service should therefore be a surgeon trained and experienced in general and orthopaedic surgery and especially in the problems of trauma'. To do this, 'the casualty departments of existing hospitals must be replanned, reorganised, and often rebuilt. There must be a separate organisation and even a separate door of entry for accident cases on the one hand, and acute medical and general surgical emergencies on the other.' The first accident service to be set up in this way was the Accident Service at the Radcliffe Infirmary, Oxford in 1941. This was run by Mr John Scott as a part time consultant, with a first assistant and two house surgeons covering the casualty department, outpatients and 60 beds.[30]

The Birmingham Accident Hospital was set up in the same year. After the war, many hospitals aspired to such a service but by the mid-1950s, it was not common. Lowden wrote: 'in a few places, where a well-organised and entirely adequate fracture service operates, the casualty department's responsibility is confined to diagnosis, except for the occasional fracture or fracture-dislocation which needs urgent release of pressure upon vessels or nerves'.[16] Lowden, a general surgeon, obviously felt that an accident service was the way forward as he said that casualty has 'frequently been required to include the treatment of ambulant bone and joint injuries and at the present time an intermediate situation is still prevalent whereby the casualty officer reduces such fractures and dislocations, then hands them on to the fracture clinic for subsequent observation and rehabilitation. This is a "Middle Ages" stage of evolution which is unsatisfactory to all concerned. If the casualty surgeon is required to deal with these injuries in the first instance, he should have the time and opportunity to follow them up and share in their subsequent treatment. If the fracture clinic does not accept responsibility for the work of the casualty department, and for the officer there, it should treat all its cases from the beginning'.[20] He appeared, however, to see the treatment of trauma as a general surgical function: 'the segregation of accident cases from other types of surgical case has many disadvantages; and as strong an argument can be produced for the development of the accident service in close association with a general surgical unit as with an orthopaedic unit'.[16] He also wrote: 'Orthopaedic problems play much less part in it [the treatment of acute injuries] than is generally supposed.'[18]

Staffing

As will be seen in subsequent pages, the Platt Report changed the name of departments and attitudes but probably had little effect on the structure and organisation of departments. Thus departments in the mid-1960s were not substantially different from those in 1960; so when describing staffing levels before Platt, I will take the liberty of including data until the mid-1960s.

Before discussing the staffing of A&E departments, it is necessary to describe medical staffing of hospitals in general as this was very different from present staffing. Until 1953, doctors registered on qualification and could, and frequently did, go straight into general practice. However many doctors did one or more house officer (HO) posts to gain further experience. On 1 January 1953, it became compulsory to do

12 months of HO posts before registration and existing posts were divided into those recognised for preregistration training and post-registration HOs. There was still no necessity to do any further hospital posts if one intended to enter in general practice but only if one intended a career in hospital medicine. The situation was complicated by most doctors having to do two years of National Service.

The Spens Committee in 1948 had recommended that there should be three training grades later named junior registrar, registrar and senior registrar[31] (SR) in which one would be expected to spend one, two and three years respectively. In 1950, junior registrars ceased to be recognised as a training grade and were renamed senior house officers (SHO). (Though interestingly in 1976, as an SHO in medicine at Guy's Hospital, I was still known as a junior registrar.) In 1951 it became obvious that the number of registrars required to provide a service exceeded the number needed to fill consultant vacancies and so that grade, too, ceased to be recognised as a training grade. At the same time senior registrar training was extended from three to four years.

At the start of the NHS, it had to absorb a number of doctors who had been practising at a senior level but who did not have the qualifications to become a consultant. These doctors were graded Senior Hospital Medical Officers (SHMO) and had security of tenure. In addition, there were some junior doctors in post HO, non-training posts but who were not of registrar status. They too had security of tenure and were absorbed into the NHS as Junior Hospital Medical Officer (JHMO). Appointments continued to be made into this grade, but often on short-term contracts.

Entry to the consultant grade was competitive and so not everyone who completed senior registrar training could expect a consultant post.[32] Those who failed could obtain an SHMO post from which they were free to continue applying for consultant posts. Many consultants, particularly in London, worked in more than one hospital. In the mid-1960s 19 per cent of London consultants worked for three or more different authorities.[33] On occasion doctors could hold posts in different hospitals at different grades, for example, sessions at one hospital as a consultant but in another as senior registrar or SHMO.

There was a great shortage of junior doctors. Hospital beds had increased from 410,829 in 1936 to 504,321 in 1950. This was associated with decreased working hours for doctors and the administrative separation of general practice from hospital medicine so that GPs no longer came into the hospitals as honorary physicians and surgeons as they had prior to the NHS. The establishment of junior posts had been expanded to cope with this. In one hospital group there had been ten junior staff below senior registrar rank in 1948 which had increased to 30 by 1953.[34] Not surprisingly, many posts were vacant. Prior to the start of compulsory pre-registration house officer (PRHO) posts in 1953, 2104 posts were approved for PRHOs in England and Wales but in 1951 only 1531 new doctors had registered and so a significant number of posts remained unfilled (though Scotland slightly overproduced doctors with 557 approved PRHO posts and 699 registrants so that 142 Scottish doctors were available for England and Wales).[35] This still potentially left 20.5 per cent of PRHO posts unfilled and in fact on 31 December 1952, 15 per cent of HO posts were unfilled with 10 per cent unfilled on 31 December 1953. A contributory factor was the low salary compared to the salary of an assistant in general practice.[36] As now, posts were more difficult to fill in non-teaching hospitals.

One of the reasons for applying for junior posts in peripheral hospitals had always been that it gave an opportunity to get acquainted with local GPs and to apply for any vacancies which became available. Around this time there was a great shortage of vacancies in general practice[37] and so these posts in peripheral hospitals became less desirable. National Service aggravated the shortage of doctors.

There was also a shortage of consultant posts associated with an oversupply of senior registrars (SRs). In 1953 there were at least 279 SRs whose appointments were shortly to be terminated but in 1952 only 25 consultant posts in general medicine and 28 in general surgery were advertised.[38] Not surprisingly emigration was high. Immigration was encouraged to fill junior posts with overseas-born doctors tending to congregate in the less popular specialties and in the Northern industrial towns.[39]

The vast majority of doctors were male, hence language such as a recommendation in the Nuffield report that casualty departments 'should have immediately available at any one time a medical man of consultant quality...'[11]

Early 1950s

Lowden in 1956 stated that initially there were three main ways of staffing casualty departments.[16] The first was with GPs attending at prearranged times or on a rota. This system (he said) often worked well.

The usual method was with a rota of the resident doctors. Often the duty House Surgeon (HS) was nominally on call for casualty but he was often the least able to attend rapidly. In practice he was often covered informally by 'extemporisations made amongst the residents themselves – residents of all specialties "rallying round" a young colleague who is under pressure of work'.[16] For its success, this depended on a spirit of fellowship among residents. However relief duty in casualty was often irksome and a cause of friction when morale was low.

Most larger general hospitals had an establishment of a casualty officer but this was frequently a shared appointment, for example, with orthopaedics, ENT, eyes, skins or anaesthetics. It was seldom an attractive post and often the last to be filled. Not everybody agreed that casualty departments should have their own staff: in 1952 one hospital had been told by the Regional Board that there should be no separate junior medical establishment for casualty.[11] A few larger non-teaching hospitals had more than one casualty officer, usually a registrar or an SHO, and a House Surgeon. Staffing levels often depended more on custom than reason with some hospitals with two casualty officers doing no more work than hospitals with one. Some hospitals had more than two, 'allowing generous off-duty periods for study'.

The only detailed description of the staffing of an individual hospital, which I have found from that era, was Guy's Hospital and this is far from typical. In 1949 the casualty department was staffed from 0900 to 1900 hr by two outpatient officers (presumably casualty officers), three dressers (medical students) and the casualty surgeon (presumably not present all the time). From 1900 to 0900 hr, it was covered by an assistant house surgeon (who also had duties elsewhere in the hospital) assisted by two dressers. Arrangements were later altered so that one outpatient officer would stay until 2100 hr.[13] The number of consultant sessions was not stated though by 1966, there were nine consultant sessions per week in the department.[40]

Whenever there is an absolute shortage of doctors, posts in unpopular specialties are the most difficult to fill. Casualty was unpopular. The only things helping to fill posts in casualty were the requirement by the Royal College of Surgeons (RCS) that applicants for the Fellowship exam had to have completed a six-month post in a casualty department and (later), immigration of doctors. There were several reasons why casualty was unpopular. Some of these are detailed in letters to the *Lancet*. There were (as now) medico-legal hazards.[41-43] A more severe problem, which exacerbated the medico-legal problem, was the lack of back up and support for inexperienced doctors. 'Theoretically advice is available; but in practice surgical registrars and others are busy with their own work. Casualty tends to be regarded as a nuisance, and the asking of advice as a mark of inefficiency.'[44] Asking advice 'is relatively easy in a teaching hospital ... but in some provincial hospitals resort has to be made to the consultant. ... A casualty officer is perfectly free to call in a consultant, he is unlikely to remain very popular ... if he calls them into the casualty department perhaps several times a day.'[43] The lack of anybody more senior in the department to turn to was a pointer that there were no career prospects in the post and this too was a further disadvantage. 'The problem here has resulted from an almost total disregard of the skill, versatility and high degree of "occupational risk" involved in this perennial professional cul-de-sac. ... It is a matter for small wonder that applicants for these posts are much too scarce. Absence of prospects plus a risk of litigation are hardly inducements. Hence the casualty officer situation approaches a crisis.'[42]

A solution was obvious. A correspondent to the *Lancet* wrote: 'I would suggest that the solution lies in making more senior appointments with more recently qualified assistants in the larger hospitals. ... There must be many ... who would take up casualty work as a career if conditions and rates of pay were made attractive.'[44]

Senior Casualty Officers (SCO)

In 1953 because of the shortage of junior doctors working in casualty, their lack of supervision and the numbers of senior registrars who could not obtain consultant posts, Dr. Patterson (Senior Medical Officer, Newcastle Regional Hospital Board) suggested appointment of senior doctors in casualty. A new grade of Senior Casualty Officer (SCO) was authorised by the Department of Health (DoH) in a circular dated 7 September 1953. They were to be paid on the SHMO pay scale but they were not SHMOs and only had a four-year contract though this could be renewed. The grade was slow to take off as shown in Table 1.3 but by 1958, 72 posts had been approved and 61 SCOs had been appointed.[11] Despite the poor career prospects in other specialties, these posts were not necessarily easy to fill. Whereas 11 posts were filled within three months, 13 took over 12 months to fill and some posts could not be filled because of lack of suitable applicants. Other hospitals did not make appointments because they did not want to make appointments outside the usual hierarchy of consultants and junior doctors and in recognition that this was not a permanent solution. It was not intended that these doctors would remain in casualty forever and it was thought that many might return to general or orthopaedic surgery.

The Nuffield Report looked at the SCO grade (described in the report as SHMOs, though technically they were not) and found that most were carrying out a good job, often in discouraging circumstances. The lack

Table 1.3 Appointments of SCOs

Year	Number of SCO appointments
1953	2
1953	5
1955	11
1956	13
1957	15
1958	15
1959	15
1960	16
1961	16
1962	16

Source: Appendix to Memorandum produced by Senior Casualty Officers Subcommittee of CCSC 1964.

of a permanent appointment was bad for morale and some were frustrated with at least one that they interviewed, wanting to return to general surgery. Lowden felt that SCO posts were usually filled by 'senior registrars who have exhausted their inpatient appointments without an early prospect of promotion, and who, for the most part, have their eyes turned elsewhere'.[16] The SCO post-holders were not expected to spend all their time in casualty. The Nuffield report comments on one SCO: 'he virtually does no surgery now. ... The waiting list for "cold" hernias which the senior casualty officer is quite capable of doing, is up to three years. This is a waste not only of surgical skill and experience but of money since the senior casualty officer is paid an SHMO salary for an "office hours" diagnostic job.' Of an SCO appointed after nine years as a surgical registrar, it says: 'he is, in fact, carrying on quite a fair surgical practice in the department, dealing with major compound fractures as a routine'.

Three surveys were done of SCOs which enable us to obtain a picture of the kind of people they were and the work they did. The Nuffield team surveyed 59 SCOs. Fifty-one were British graduates, four were Commonwealth graduates and four qualified elsewhere. Thirty had either an Fellow of the Royal College of Surgeons (FRCS) or an MD.

In March 1962, Lamont (a SCO) sent out a questionnaire to 71 of his colleagues and got 47 replies from SCOs who had been in post between three months and nine years.[45] Six were in their second SCO post. Most were true SCOs and had a four-year contract but a few were SHMOs and thus had a permanent post (these were mostly in Scotland).[46] Their mean age was 40. Forty-one were British Nationals and the rest were mostly from the Commonwealth. Ten had a higher qualification and eight had been senior registrars for between one and six years. Forty four spent between 20 and 100 per cent of their time in the casualty department. Most spent the rest of time in general surgery and orthopaedics. Only three spent more than 50 per cent of their time in 'specialties like medicine, obstetrics and anaesthetics'.

The SCOs were asked for their preferences for a consultant post, the answers are given in Table 1.4.

The Casualty Surgeons Association (CSA) archive also contains data from another survey of SCOs. This is undated but probably occurred in 1963 or 1964. Sixty-six SCOs responded to a questionnaire and 27 of their responses were analysed. Only three spent all their time in casualty, ten spent 50–80 per cent of their time in casualty and 14 less than 50 per cent of their time in casualty. In the time spent outside casualty, 22 did sessions in orthopaedics, 14 did sessions in orthopaedics with some spending time in casualty, general surgery and orthopaedics. Six had been SRs.

Those who had been SRs could obtain consultant posts but this was uncommon and only ten ever became consultants, all in general surgery. Two of those who became consultants continued to have an interest in casualty work.[47]

Most SCOs were probably used as experienced pairs of hands to cover the day-to-day work of the departments, supervision of more junior doctors and teaching but with no responsibility for planning services. However some were members of their Hospital Medical Committee and so, presumably, had additional responsibilities. Of 49 SCOs who answered the Nuffield Study survey, 18 were members of the Medical Committee but 31 were not. In 1959 SHMOs who were recognised as

Table 1.4 Preferred career specialty for SCOs 1962

Specialty	Number of SCOs
Casualty or trauma	18
Orthopaedics	6
General Surgery	12
Medicine	1
Undecided or not interested	10

Source: CSA/BAEM Archive.

carrying consultant responsibility were given an extra allowance of £550 per annum in recognition of this responsibility.[31] This was refused to SCOs because casualty was not a specialty and therefore there could not be consultants in casualty.[48]

Staffing of departments in the late 1950s and 1960s

The staffing of a few hospitals has been described in the literature. Data from a number of papers is summarised in Table 1.5. These show much variation and the pattern is little different from the early 1950s. About 60 or 70 departments had a SCO as noted above. Teaching hospitals made much use of medical students who had service role for preliminary clerking of patients, assisting with dressings and doing practical procedures as well as being taught.

The Nuffield study investigated the most senior person working in the casualty department in those hospitals which did not have SCOs. The results are shown in Table 1.6.

The Nuffield team also tried to assess the quality of different aspects of the casualty department in the 18 hospitals they visited. This grading was based both on objective data on staffing levels and so on and also on more subjective data on the quality of the doctors they met, quality of service which they saw being provided etc. The results of their assessment of the quality of casualty officers are shown in Figure 1.1. (These figures are drawn using the Nuffield data: the figures do not appear in their report.) The visiting team also commented on a marked difference in quality between the holders of posts in teaching hospitals and in other hospitals.

Although many departments had had full time casualty officers during working hours they often relied on other doctors covering the department outside these hours. The doctors involved in covering the department are shown in Table 1.7.

The Nuffield team's assessment of the quality of the out of hours cover is shown in Figure 1.2. Their conclusion was that '...almost all the arrangements for night and weekend cover were patchy and inadequate. These duties are performed by house staff – often unwillingly, which does not make for effective cover.' In one hospital there was such unwillingness that there had been two mutinies by junior staff.

The difficulty in filling posts in the early 1950s has been mentioned above. The situation was no different by 1958. The Nuffield Report notes that posts were often difficult to fill and a large number of posts were filled by overseas graduates. In a survey of 52 departments in 1958,

Table 1.5 Staffing levels in A&E departments

Year	Hospital	New patients	Total patients	Staffing	Cover	Reference
1941	Radcliffe Infirmary			Part time consultant First assistant 2 HS	Also cover 60 orthopaedic beds	Robb-Smith[30]
1949	Guys				0900–1900 hrs 2 Outpatient Officers 3 dressers (medical students) casualty surgeon 1900–0900 hrs 1 assistant HS 2 dressers	Clarkson[13]
1958		23,000	35,000	1 SHO	Works 0900–1700 hrs weekdays 0900–1300 Saturdays Nights and weekends cover by HP and HS	Nuffield[11]
1958				2 SHO	Both work 0900–1300 hrs Between them cover all hours except Sat 1730–Sun 0900 hrs and Sun 1300–Mon 0900 hrs	Nuffield[11]
1958				1 SHO	Works 0900–1800 hrs night cover by rota of 4 HOs Weekend cover by Resident surgical officer (RSO) and ortho reg	Nuffield[11]

Year	Hospital	Number	Staff	Reference
1958			4 SHO 2 GPs doing evening sessions twice a week	Nuffield[11]
1958	Kings College	29,000+	2 post reg HO 3 pre reg HO (2 shared with other depts)	Fry[49]
1964/65	Royal Aberdeen Hospital for Sick Children	8,816		Surgical registrar during day House officer evening and night	Jones[50]
1965	Queen Marys, Sidcup	12,600	1 casualty officer Duty HO 2000–0900 hrs	Tatham[51]
1965	Preston RI	25,800	1 SHMO 1 JHMO 4 SHO	At least 2 on duty all hours of day and night	Garden[52]
1965	Bradford	39,714	Cons Ortho surgeon (3 sessions) 5 SHOs (shared with ortho) 11 Clinical assistant (CA) sessions		Naylor[53]
1966	Guys	76,000	5 casualty officers plus HOs 9 consultant sessions	CO work 0900–2000 hrs with half day alternate days At all times 2 CO seeing emergencies 1 seeing ambulatory accidents and infections	Clarkson[40]

Table 1.5 Continued

Year	Hospital	New patients	Total patients	Staffing	Cover	Reference
1970	Belfast	57,200	87,000	Medical assistant 2 Registrar 2 SHO 2 HO 4	HOs work rota over 24 hrs SHOs and reg work 0900–1800 hrs and 1800–2200 hrs 4 days a week	Wilson[54]
1971/72	Hereford	16,000		1 medical assistant 3 SHOs		Hardy[55]
1972	Stoke			1 orthopaedic SR (in accident unit) 3 SHOs (rotate with orthopaedics) 2 clinical assistants working sessions (number not stated)		Wainwright[56]
1972	Bradford			Cons. Ortho sessions 6 SHOs (with ortho) 16 CA sessions	3 in morning 2 until 2300 hrs 1 2300–0900 hrs	Naylor[57]
1972	Portsmouth	37,284	53,110		1 MA and 1 SHO before noon 1 doctor in afternoon 1 doctor in evening	Denham[58]

Table 1.6 Grade of most senior casualty post in departments without an SCO 1958

Grade	Number
JHMO	17
Registrar	9
SHO	23
Part time SHO	1
House-surgeon	1
GP	1

Source: Nuffield Report.

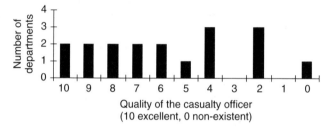

Figure 1.1 Quality of the casualty officer

10 represents a first class, keen SCO; 9 represents a satisfactory SCO, 7–8 represent a keen SHO using the post as a proper stepping stone to a surgical career; 3–6 represent less good SHOs or SCO or a conscientious HO; 2 represents a poor SHO, 1 is below a reasonable standard of quality and 0 represents severe staffing problems such as the department is covered by a succession of locums.

24 had at least one unfilled post during part of the year with four departments having a post vacant for more than 20 weeks during the year. Vacancies were filled with locums, a rota of house officers or both with general practitioners occasionally being used. They comment that 'practically the only hospitals which had little difficulty in getting junior

Table 1.7 Rota for out of hours cover included

Grade of doctors	Number of doctors
Casualty Officers giving continuous cover	13
JHMO	3
Registrars	9
House Officer only	16
Pre-reg HO	7
GPs	2
SHOs	5

Source: Nuffield Report.

Figure 1.2 Quality of out-of-hours cover

10 indicates that the department is fully covered by casualty officers with consultant advice readily available. 7–9 indicate 24-hour cover by casualty officers above HO grade and 3–6 that the department was covered without difficulty by a rota of junior house staff or GPs. 2 indicates the above level of cover, achieved with difficulty. 1 indicates that there was great difficulty in organising out of hours cover which caused bad relations among junior staff and 0 represents totally inadequate arrangements.

Table 1.8 Most senior doctor seeing new patients

Grade	%
HO	23
SHO	34
JHMO	9
Registrar	8
GP	7
Senior registrar	1
SHMO or equivalent	7
Consultant	2
Non-medical staff	8
Unknown	1

Source: Platt Report.

staff were teaching hospitals, or those non-teaching hospitals which had reasonably up to date accommodation and consultant cover'.

Further information on staffing can be obtained from the Platt Report.[1] For one week in October 1960 a survey of the 789 casualty departments was undertaken. Among other questions, they asked for the rank of the most senior doctor seeing new patients. The results are shown in Table 1.8.

There was much variation in different types of hospital and in different parts of the country. In teaching hospitals, very junior doctors saw a higher proportion of patients. In one London teaching hospital 95 per cent of new patients were seen by a HO and in a provincial teaching hospital 98 per cent were seen by either a HO or an SHO. However there were also two non-teaching hospitals where over 90 per cent of new patients were seen by a HO or SHO. Platt reports that '...at many hospitals recently qualified staff when serving in the casualty department undertake responsibility which they would not have on the wards or out-patient clinics'.

The proportion of new patients seen by a general practitioner varied. In teaching hospitals it was only 2 per cent except in Leeds where 20 per cent of patients were seen by a GP. Apart from that city, the highest proportions of patients seen by a GP were in the South West (18 per cent) and Wales (14 per cent). The Nuffield study reported that general practitioners were rarely used. This may be largely because of the expense. A general practitioner was paid 3–3½ guineas for working a single 3½ hour session whereas a house surgeon was paid £500 per year.[11] Whereas general practitioners were probably usually used to cover the new patient workload, they were sometimes used in a supervisory and training capacity.

Consultant involvement

Consultant involvement in most casualty departments was minimal. Clarkson reports that departments were commonly the 'nominal responsibility of one of the consultant staff, generally the senior general surgeon who may or may not visit it once or more a year'.[21] Sir Robert Platt agreed by saying that in most departments 'consultant supervision is insufficient and in some, it is purely nominal'.[31] Sir Harry Platt had noted that: 'some Regional Hospital Boards have made a number of consultant appointments with contracts including duties in casualty departments. Over the country as a whole however, formal arrangements of this kind are rare.'[1]

Table 1.8 shows that in the Platt survey, only 2 per cent of new patients were seen by a consultant and many of them had been seen by a more junior doctor as well. Some consultants did a weekly or twice weekly clinic in the department but they saw patients with their special interest, for example, septic hands or fractures rather than patients according to the severity or length of illness. There were exceptions. At Guy's Hospital, Patrick Clarkson, who had a distinguished war service as a plastic surgeon, had been appointed consultant in charge of casualty after the war and there were nine consultant sessions per week allocated to casualty with about half of them being done by him. There were daily sessions of between one and four hours.[40] He appears to have taken a very active involvement in the department if the week ending

4 November 1960 is typical as during that week 10 per cent of the new patients were seen by a consultant and 50 per cent of the follow ups were seen by a consultant or registrar.[59] He was very enlightened for the time in believing that casualty officers needed at least a weekly seminar at which current work and results could be closely reviewed and topics of interest to casualty work discussed.

Departments outside London also had consultants with sessional commitment to casualty. Garden, describing the casualty department in Preston states that the futility of staffing the department by a rota of newly qualified resident officers was recognised before the introduction of the NHS, and a fully qualified junior surgeon was appointed to work in the department.[52]

Lowden, a surgeon in Sunderland had a major interest in the casualty department of which he was in charge and wrote several articles (some of which have been quoted above) on the subject. He also wrote probably the first textbook of casualty[20] which he described as 'the result of a seven-years' study in the casualty department of a medium-sized, non-teaching provincial hospital' and the many case histories which he describes show his day-to-day involvement in the department.

In 1949 at Leeds General Infirmary, Maurice Ellis was appointed a senior registrar in the casualty department and in 1952 was appointed the first full time casualty consultant.[45]

At the other extreme, Clarkson reports that in at least one teaching hospital in London the department was run by an SHO or junior registrar responsible only to the lay authority of the hospital.[21]

In two surveys of departments, the Nuffield team asked who was in charge. The results are shown in Table 1.9 (It is not stated whether the

Table 1.9 Who was in charge? – Two surveys 1958

Specialty	Number of departments	
	First survey	Second survey
General surgeon	16	15
Orthopaedic surgeon	31	23
General and orthopaedic surgeon jointly	8	4
Medical superintendent	1	2
No consultant's responsibility	4	12
Medical SHMO		1
Surgical SHMO		2
Information not given	8	

Source: Nuffield Report.

departments questioned in the two surveys were mutually exclusive and so I present them separately). Sometimes there was uncertainty. At one hospital the management stated that the consultant orthopaedic surgeons were mainly responsible for the department with the senior surgical registrar dealing with the day-to-day medical staffing. However, when the hospital was later visited, the visiting team were told that it was the responsibility of one of the consultant general surgeons. Sometimes consultants other than the person nominally in charge took an interest: 'interest in the orthopaedic aspect of the work has now started to grow as a result of the appointment of a second consultant orthopaedic surgeon who runs the fracture clinic and is watching the standards of the initial treatment of fractures. He has, however, no official responsibility for the casualty department.'

The study confirms the variability of consultant cover for the department. Some consultants took a major interest. 'The consultant in charge is probably unique in the active interest he takes. ... Apart from his frequent informal visits to the department he takes 2½ sessions per week actually at the same table as the SHO casualty officer seeing both new and old patients...' and for another consultant: 'when the SHO is off duty, the general practitioners provide cover for their own patients ... and if they are unable to come to the hospital quickly or a patient arrives who has no local doctor, the consultant is on second call. He does not consider it beneath his dignity to be called in at any time of the day or night to treat a patient, even with only a minor injury, and during his annual leave he frequently calls to the hospital, not forgetting the casualty department, to keep a friendly eye on what is going on.'

This study also confirms the nominal cover provided by some consultants: 'on his own evidence [the consultant's] appearance in and contact with the department is confined to taking his afternoon cup of tea with the casualty sister in her office' and, for another department: '... judging from his surprise at some aspects of the department's activities at the time of the team's visit, it would appear his actual appearances in the department are few and far between'.

By 1966, only three of the twelve London teaching hospitals had allocated consultant sessions in casualty.[40]

Even if a consultant in charge did not take an active interest, there were other senior doctors who might. Table 1.9 shows that in some hospitals the medical superintendent was in charge. (The medical superintendent had an administrative role in running the hospital but was also responsible for staff health and casualty.) A history of St Mary's Hospital (the last London teaching hospital to appoint a consultant), says that 'until

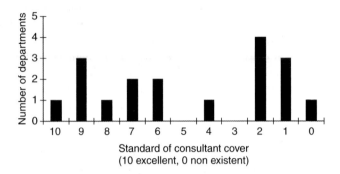

Figure 1.3 Standard of consultant cover

10 represents a full time consultant in charge, 8 and 9 represent a consultant in more than nominal charge, 3–7 represent consultants who attend for a weekly clinic but who take no substantial part in the running of the department. 2 represents a consultant who visits socially, 1 a consultant who never visits and 0 a department without a consultant even nominally in charge.

the 1980s the department was supervised by the orthopaedic surgeons but largely run by nurses and students, with Cockburn [medical superintendent, appointed in 1955] a mediator.'[60] In some hospitals this role was taken by a much more junior doctor, the Resident Medical Officer (RMO) who was usually the equivalent of a senior SHO or junior registrar. Howard Baderman and David Williams who were both to play a major role in the development of A&E, were RMOs at University College Hospital (UCH) and the Middlesex Hospital respectively.[61]

In their report, the Nuffield team graded the quality of consultant cover. The results are shown in Figure 1.3.

Among the best casualty departments were those run as accident services with the casualty department integrated with the fracture clinic and inpatient orthopaedic trauma beds with common staffing so that it was staffed by orthopaedic junior staff either on a rota or else on a rotation. The Birmingham Accident Hospital was in a city with other hospitals to which patients with non-traumatic problems could be taken but the accident service at the Radcliffe Infirmary, Oxford also dealt almost exclusively with trauma with 95 per cent of patients attending following an injury. The 1941 staffing is given above and by 1967 this had increased to 5 SHOs, 3 registrars, 2 SRs and 12 consultant sessions from 3 orthopaedic surgeons and 1 session from a consultant in plastic and maxillo-facial surgery.[62] This department had active consultant involvement, worked well and had no difficulty in attracting resident doctors: 'Posts were nearly always overapplied for without advertisement...'[63]

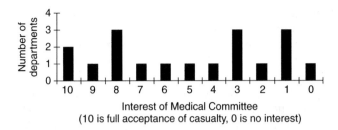

Figure 1.4 Interest of Medical Committee

10 indicates that the hospital accepted the casualty department as an important and integral part of the hospital. 8–9 indicate that either that there was an active consultant in charge (who would automatically be part of the hospital medical committee) or that the Senior Casualty Officer was a member of the Committee. 5–7 indicate that casualty had a central role in the hospital but the rest of the hospital staff were neutral. 2–4 indicate that relationships were only good between junior medical staff and nursing staff. Lip service only scored 1 and 0 indicate that casualty had a very poor and isolated position within the hospital and that no interest appeared to be taken in it.

In Bradford in the early 1960's, the accident service was staffed by three consultants, one of whom worked three sessions in A&E; 1 SR, 3 registrars, 5 SHOs and 11 GP clinical assistant sessions in the A&E department.[53] Where there was active consultant involvement, this type of department worked well. In Luton, the reorganisation of the accident department to this way of working led to the proportion of patients seen by a senior doctor to go from 10 per cent to 55 per cent.[64]

One of the major reasons why casualty departments had problems is that the rest of the hospital took very little interest in them. The Nuffield Report looked at the interest taken in casualty by the Hospital Medical Committee. The results are shown in Figure 1.4.

Summary

Before the Platt Report, the staffing of casualty departments was very variable. There were examples of excellence and not just in Leeds which was the only department with a full time casualty consultant. However consultant supervision (where it existed) was often little more than nominal. Many of the larger departments had SCOs who worked well but most were not given adequate responsibility for their departments. Junior staff was difficult to obtain and many departments relied on inexperienced preregistration HOs.

Factual data is clearly important but the true meaning of these figures can only be understood with descriptive language. In 1961 Sir Robert Platt stated: 'In our experience, the medical staffing of casualty departments is the least satisfactory feature in the present hospital staffing system.'[31] In the same year the *BMJ* in an editorial on the shortage of doctors, falling medical student numbers and high emigration rates quoted Lord Taylor in the House of Lords: 'I cannot recommend your Lordships to go into such hospitals [non-teaching hospitals] as a casualty, for there is in many cases no casualty officer. A house-surgeon will have to leave the theatre when he can, to treat you, and his experience will be far less than that of your own general practitioner. When he comes he will probably not be a British Graduate and he could well have difficulty in understanding what you say...'[65]

2
Who Should Run A&E Departments?

Reports

The severity of the problems in casualty in the late 1950s is indicated by the fact that between 1959 and 1962 there were at least four reports which made recommendations on casualty departments.

In 1959 the British Orthopaedic Association published a Memorandum on Accident Services which sought to replace casualty departments with accident departments. The BOA detailed failures of organisation, staffing, accommodation and surgical training. Too many hospitals were trying to provide an accident service with limited facilities and staffing and they recommended that services should be concentrated on a smaller number of hospitals fully equipped and staffed to provide a full service round the clock. They believed that accident services should be organised on a nation-wide scale as a quasi military operation with the country divided into areas, each with its comprehensive accident unit based on a district general hospital (DGH).[1] 'Emergency receiving rooms for other than accident cases will still be needed in designated hospitals, but casualty departments as they used to be known should no longer exist.'

The report suffered from being too focussed on trauma and essentially ignoring other emergencies. The *Lancet* agreed that improvement was needed but urged caution,[2] believing that too much emphasis had been placed on the well-equipped and comprehensive centre and that 'limelight on the specialised centres must not cast a shadow over the work done in collecting and screening the casualties on their way'. It reasonably asked what was major and what was minor and said that surely an inhaled foreign body is an accident. It felt that 'a case can be made out, in fact, for a review of the arrangements for hospital treatment of all emergencies' and asked: 'if casualty departments...are to exist no longer, what is to replace them? Where shall we send the overdose, the patient with bleeding varicose veins, the patient rescued from a gas filled room or the paraphimosis?'

The *BMJ* felt that organising a service by the state as a quasi-military operation did not seem appropriate but that each region should develop its own accident service according to its needs[3] but others in the correspondence columns of the *BMJ* were more critical of this report: 'it would be quite disastrous to abolish these departments without providing alternative services for the many non-traumatic conditions that are seen by the young casualty officers. The role of the casualty department must be reviewed.'[4]

In 1958 the Nuffield Provincial Hospitals Trust commissioned a report after hearing serious concerns about casualty services up and down the country. This report, which I call the Nuffield Report, was published in 1960. It made recommendations purely on casualty departments.[5] They recommended that '... there is a need for leadership' and 'the medical staffing of such services demands special attention, particularly the provision of adequate consultant cover ...' They asked 'for the real, as distinct from nominal consultant cover in the main casualty departments' and felt that 'it is irrelevant what kind of specialty such men should have. They should, however have a broad knowledge and be trained to recognise those conditions in need of urgent treatment.' They felt that the best way to organise this was 'a mixture of, say, a new part-time appointment for a "casualty" consultant covering six or seven sessions a week, the balance of his time being covered by requirements written into existing surgical contracts, might be the right line to follow'. They also felt that the ideal casualty department 'should have immediately available at any one time a medical man of consultant quality...'

In 1961 the Accident Services Review Committee produced a report which, like the BOA memorandum, concentrated entirely on accidents.[6] It recommended a three-tier system of central accident unit, based on a teaching hospital; accident unit in a district hospital and peripheral casualty service, based on cottage hospitals, GPs' surgeries and industrial medical centres. The accident unit should be staffed by general and orthopaedic surgeons.

In 1962 Sir Harry Platt produced his report which had been commissioned by the Department of Health and whose terms of reference were 'to consider the organisation of hospital casualty and accident services and to make recommendations regarding their future development.'[7] As part of the report, a study was done of all the patients seen in the country during a single week in October 1960. This found that trauma made up 61 per cent of the workload, medicine 9.3 per cent and surgery 7.9 per cent.

The report made 45 recommendations. It rejected the three-tier system in favour of a two-tier system of accident units in larger hospitals with smaller hospitals only treating patients with minor injuries. Like the BOA and the Accident Services Review Committee it recommended that 'the number of accident and emergency units should be greatly reduced, so that each can be adequately staffed at all times. A unit should not normally serve a population of less than 150,000.'

On staffing, Platt recommended, on the basis of the predominance of trauma, that 'the medical staffing of major accident and emergency units should be increased to allow each unit to have three consultant surgeons, each devoting a substantial part of his time to this work, supported by adequate numbers of intermediate and junior medical staff.' At least two of these should be orthopaedic surgeons. He also recommended that it was 'necessary to appoint one consultant to be in administrative charge of an accident department. ... By far the greatest part of accident work falls within the province of the orthopaedic surgeon. It is therefore normally the best arrangement for a senior orthopaedic surgeon to have day to day control of the accident and emergency department. ... Policy ... should be decided by a small team representing each of the specialties directly concerned with the accident service.' However on examining the figures, most of the trauma was minor and only 13.1 per cent of the workload consisted of fractures, dislocations and multiple injuries (which may or may not have represented orthopaedic trauma).

He recommended that the service needed three officers of intermediate grade and an adequate number of SHOs to give a service at all hours of the day and night. The intermediate grade doctors could be either registrars on rotation or doctors in the grade of Medical Assistant (MA). He felt that GPs could make a real contribution to the running of a department.

Another recommendation was that the word 'Casualty' should be abandoned as the description of any type of hospital service, department, officer or patient and should be replaced by 'Accident and Emergency'. 'Accident is put before emergency because the number of accident cases is far greater than the number of other emergency cases reaching the department. It is misleading to think of accidents and emergencies as two streams; one is a river and the other a trickle.' He rejected the term 'Accident service' feeling that this might cause confusion in deciding where to take the person who collapses in street but in the report, the differences between A&E departments and the accident service is frequently muddled. Unfortunately, as Rutherford (A&E consultant in Belfast) pointed out, 'the phrase "accident and emergency services" is sometimes used [in the report] to denote the work of the casualty department, sometimes to denote accident services and sometimes to include both.'[8] Rutherford also noted that 'this confusing nomenclature has continued in many subsequent publications'.

Platt also made various recommendations on the design and organisation of A&E and recommendations on the orthopaedic accident service and ambulance service which need not concern us in this book.

Apart from these four reports dealing with accident and casualty services, Sir Robert Platt in his report on medical staffing[9] recognised a requirement for consultant involvement but did not appear to feel that it needed much time: 'the first principle in the staffing of casualty departments should be that one or more consultants should have a specified responsibility for casualty work, and a definite and sufficient part of their time allotted to its supervision. Much of the work is of a kind in which a policy for treatment can be defined reasonably closely and supervised by the senior staff concerned. The seniors would be available to assist with difficulties, and should have regular times of attendance at the casualty departments, though this should not become a substitute for ordinary consultative outpatient work.'

Maurice Ellis had been appointed the first full time casualty consultant in Leeds in 1952. Here he devoted his energies to running a very good department and organising teaching and research rather than becoming involved nationally with the organisation of casualty services. The first person to argue in the medical press that full time casualty consultants was the best way of managing casualty departments appears to have been Lamont, an SCO in Grimsby. In 1961, after painting a picture of the variety of work and the inexperience of the doctors, he argued the need for 'an integrated, self-contained casualty service, preferably on a national basis, in which the acolytes can be trained for the subtleties of the work'.[10] He felt that each casualty unit should have at its head a consultant of high attainment – ideally a man with surgical training and qualifications in industrial medicine whose 'primary function ... is to be a father figure to the young men who will make their career in the casualty service – translating into reality the platitude that casualty work can provide the young with invaluable experience'. He recognised that '... specious talk of "under the supervision of" or "under the general direction of" extraneous consultants will not do' and that the person in charge needed to be on the spot, giving most of his time to the department. He needed to be a consultant for the authority and to protect his juniors against unfair criticism. This person would be a traumatic surgeon who did not need to be an expert in all branches of surgery but must be cognisant of them all. They would have a close alliance with one branch of surgery and will usually need to take a special interest in brain injury. Their role was 'to document and put the patient in a state of preparedness for the special skills of the appropriate department'.

In letters to the *Lancet*, Lamont's views were supported by Maurice Ellis who felt that the existing SCOs should be given consultant status,[11] but were not popular with those who felt that casualty departments should be integrated into the hospital and managed by committee. '... To have a director of accident services... is the surest way to wreck the harmony which is so essential in any hospital between consultants. I feel that if anybody is to direct an accident unit, it ought to be an orthopaedic surgeon... who would do the job on an annual basis, and rotate with his colleagues.'[12]

The Ministry of Health accepted the Platt Report and Lamont, with immense optimism, welcomed the news that 'the outmoded system of placing conscientious but not specifically trained juniors in charge will cease; we are to see not less than three surgeons of consultant status, giving 24-hour cover where populations of 15,0000 or more are at risk'.[13] The *Lancet* felt that the Platt Report was excellent but that it lacked a sense of urgency. It also felt that a change of name might not do much to abate the problem of casual attenders.[14]

Concerns of the Senior Casualty Officers

A flaw in the Platt recommendations was to assume that inclusion of duties in a contract of an existing consultant was synonymous with interest and conscientious supervision by the person concerned and G. da Costa (SCO, Bishop Auckland) asked in a letter to the *BMJ*: 'Would it not be better... to at least throw the doors open for the attainment of consultant status to such of those persons who are genuinely interested in making careers in this field?'[15]

The SCOs had other objections to the Platt Report which they expressed at a meeting in 1963. They felt that as the only senior doctors devoting the majority of their time to casualty, they had a valid view on the future of their specialty and they were concerned (probably infuriated) that Platt had not made an approach to draw on their experience, or obtain their opinions.[16]

They felt that the Platt Report had given 'insufficient attention...to develop the 24-hour Emergency Services (including Trauma) as an integrated organization under whole time personal management' and that 'hospitals do or will need people who are willing to undertake this work as a full time interest'.[17] They felt that these should be consultants in 'Casualty (Traumatic and Emergency) work'. They also felt that Platt had concentrated too much on the management of trauma and had given insufficient emphasis to what they called the 'Emergency Situation': '...giving medical aid in an emergency is the most important of the functions of Casualty Departments'.[16]

The SCOs also felt threatened. Under the recommendations of the Sir Harry Platt report, there seemed little place for specialists in casualty or pure A&E work. In addition the Sir Robert Platt report had recommended abolishing the SHMO grade and replacing it with a new grade of Medical Assistant (MA) who 'should work as assistants to consultants and under their supervision'. The report also recommended that 'all patients in the Hospital Service should be in the charge of consultants, and consultants in the specialty or specialties involved in the treatment of each patient should bear the responsibility for his care and the medical work required for him'. MAs, while still experienced doctors, would be considered more junior to an SHMO and have a lower pay-scale. The report had also recommended that '...the Medical Assistant grade should help with the staffing of casualty departments. ...Where appointed Medical Assistants should be fully embodied in the "team" and care taken to maintain their interest in the development of the service'. MAs should also only be appointed if 'a satisfactory standard of consultant staffing is provided'. It appeared to the SCOs that there was no satisfactory consultant staffing in casualty and that if this report was taken at face value, orthopaedic surgeons could not take responsibility for anything other than trauma and that casualty departments would cease. Pascall, an SCO in Plymouth pointed out in a letter to the *Lancet* that an orthopaedic surgeon working sessions alongside a SCO would be an amateur in the medical, surgical and social problems which arise.[18]

Senior Casualty Officers Subcommittee

The SCOs needed a forum to meet to ensure that they were speaking with one voice and an organisation to represent their interests. In August 1962 Mr G. da Costa asked to form a committee within the British Medical Association (BMA) as they could not be represented by either the Junior Doctors Group or by the SHMOs. On 4 March 1963 and the morning of 5th, 13 SCOs who had a dominant interest in casualty departments met in BMA House to discuss the Platt Report and the serious situation which would occur if the reports were implemented. Mr da Costa was elected chairman of the group and Edward Abson, an SCO in Southampton, was elected secretary and treasurer. In the afternoon they met a group from the Central Consultants and Specialists Committee (CCSC) of the BMA led by Mr H. Langston an orthopaedic surgeon and, incidentally, Mr Abson's consultant. The CCSC agreed to help them and recommended that they formed a Senior Casualty Officers Subcommittee of the CCSC. The SCOs met again and agreed, electing eight representatives.[19] The Council of the BMA approved and the first meeting of the Subcommittee consisting of five SCOs and three members of the CCSC was on 27 June 1963.

The BMA was happy to fight for the problems which the SCOs were having but pointed out that the appointment of full time consultants in casualty would mean a complete reversal of Department of Health policy and was contrary to the recommendations of Sir Harry Platt who was a highly respected elder statesman of medicine. To achieve this required evidence and the committee resolved to write a memorandum to argue the case. Mr da Costa wrote the first draft and after a number of meetings and despite feelings, at times, of 'being sidetracked by delaying tactics into "procedural" channels' and that Mr Lewin, a neurosurgeon and chairman of the Subcommittee, 'was blinded a little in his confidence in his own knowledge of casualty work',[20] the memorandum was prepared and presented to the CCSC in March 1964. In summary it stated:

1. The Platt Report had overemphasised the traumatic elements of casualty work. Better provision for the management of the 'Emergency Situation' from whatever cause should also be made.
2. Evidence was presented that experienced doctors such as SCOs had a role to play in continuing the development of A&E services.
3. A career grade of SCO leading up to consultant level was proposed and a proposed plan for training in the future was detailed. It was recommended that some suitable SCOs could immediately be up-graded to consultants. They believed that 28 of 66 SCOs would be suitably qualified and experienced to be upgraded.
4. They felt that their proposals were complementary to those of the Platt Report. Accident centres were still required but for the full development of A&E services to cater for all needs, there was a place for men who would make this work their primary interest.

In 1964 the SCOs' fears were reinforced by the Ministry of Health's response to the Sir Robert Platt report. SHMOs would be absorbed into the MA grade, keeping their existing salaries. The response stated that future appointments as SCO should be as MAs and that hospital boards 'should consider how far existing senior casualty officers, who so desire, can be appointed to posts in the new grade'.[21] SCOs were in a different situation from SHMOs in that they only had a four-year contract, there was no right of transfer and there was the possibility of losing their jobs. This was not just a theoretical possibility as by March 1965 it was pointed out to Mr Lewin in a letter that at least one SCO had been notified that his appointment would be terminated and that no fresh appointment will be made, the work being done by an orthopaedic senior registrar.[22] Even if they did transfer to the MA grade they stood to take a drop in salary.

This memorandum was passed to the Orthopaedic Group Committee. They responded that 'the suggestion of a career grade in casualty, accident and emergency up to consultant level is not accepted. There should be no consultant working solely in a casualty department.'[23] It is worth noting the orthopaedic surgeons' objections and the SCO responses[24] in some detail as it is important to understand orthopaedic objections.

- A career grade in casualty would deprive all other surgical units of the stimulus of emergency surgery.
    Response: One casualty consultant will not be able to do all emergency surgery and, if in a large department, probably will not have time to do much surgery at all except for any special interest.
- A casualty consultant invariably tends to set up a minor orthopaedic service and retains cases which should be referred.
    Response: Only one casualty consultant exists, so conclusions cannot be drawn.
- There would be difficulties in relationships between orthopaedics and casualty.
    Response: This need not happen.
- The orthopaedic surgeons stressed the importance of separating out major and minor casualties.
    Response: It is important to avoid the duplication of services and major trauma may present with minor symptoms.
- A senior registrar should be in charge and a consultant could supervise for two sessions per week.
    Response: This was not in accordance with the Sir Robert Platt Report which said that a consultant should be the cornerstone of any service.

- Casualty should be in charge of a committee.
  Response: It would not work in practice.

Other arguments against consultants in casualty put forward in various articles and letters were:

- That casualty was not a specialty. Casualty consultant 'is a double misnomer; first it implies that casualty work is a specialty, and, secondly, that there exists a type of specialist in this work'.[25] 'One of the common objections to appointing consultant casualty officers has been that trauma involves more than one specialty.'[26]
- That orthopaedic surgeons could (and should) manage the trauma and that it was not necessary or desirable to provide a consultant service for casual patients.[27]
- That a single A&E consultant could not make a difference to a 24-hour service. 'I cannot see how a casualty surgeon appointed to oversee an emergency department working 9–5 and not weekends can play any useful part in the department whatever.'[28]
- Orthopaedic surgeons had had to fight to take over the care of trauma from general surgeons but had dramatically improved the care of patients with fractures. One of the principles that this improvement had been based on was continuity of care. Orthopaedic surgeons feared that A&E consultants doing the initial management of fractures would break this continuity of care and (even worse) might not refer such patients on to orthopaedic surgeons. Hence there were statements such as: 'there is no place for a consultant in the accident and emergency department who has no responsibility for the continuing care of the injured patient'[29] and 'orthopaedic surgeons... should insist that the management of locomotor injuries *from the outset* should remain their responsibility. [To do otherwise] he will be doing a grave disservice, not only to his patients but to his forebears who fought for the development of fracture clinics and the first-line care of injured patients during and after two world wars.'[30]
- There was a belief that a post without inpatient beds and without continuity of care would be unsatisfying: '... Most of these men have no wish to be casualty surgeons. They desire a post in one of the more definitive specialties. Apart from those who would take any kind of post rather than return to their respective native lands, few would obtain real job satisfaction in such posts.'[31] '... Even when casualty is a direct consultant responsibility, the consultant cannot be expected to spend the whole day there: the range of clinical work is too limited.'[32]

- Finally, when it appeared that A&E consultants were inevitable, came the argument that there were too few people suitably qualified to take these posts.[29]

It is quite clear from all the literature that both the protagonists and antagonists of the idea of casualty consultants saw such consultants as having a surgical training. However the SCOs did not, necessarily, wish to operate and saw the work as more than trauma realising that poisonings, medical emergencies, psychiatric problems and social emergencies caused more problems than fractures.[33]

The memorandum was then discussed by the Joint Consultants Committee (JCC) who sympathised but did little.

It needs to be said that language complicated the understanding of this intended new specialty and provoked unnecessary fears among orthopaedic surgeons. As far as most A&E doctors are concerned, the Platt Report recommended renaming the 'casualty department' the 'A&E department' and the words casualty and A&E are synonymous. (This is not to say that individual A&E doctors did not have strong – sometimes very strong – views for one or the other.) To orthopaedic surgeons, 'casualty consultant' and, even worse, 'casualty surgeon' implied that the new specialty would be 'casualty surgery'. The BOA memorandum of 1959 had said: 'the creation of "casualty surgeons" who would accept full responsibility for the treatment of injuries in every part of the body would therefore be an undesirable development'[1] and even orthopaedic surgeons such as Garden who supported the idea of consultants in casualty, opposed this idea.[34] An example of this misunderstanding is shown in a BOA resolution in 1971: 'we do not support the creation of casualty consultants, we do support the view that there should be a consultant in accident and emergency medicine...'[29] though, as will be seen later, their concept of an A&E consultant was very different from that which most people understand. To call their association the 'Casualty Surgeons Association' was not a very good political move for the SCOs.

One suggestion made by the BMA had been to offer the SCOs 'rehabilitation' by becoming orthopaedic senior registrars but while this was a possibility for some of the recently appointed SCOs, it was obviously not an option for those in post for ten years or more.[35]

The creation of consultants in casualty and the predicament of the SCOs were separate issues. The second problem was solved in 1965 with the SCOs obtaining job security and the right to transfer to the MA grade without detriment to their remuneration.[36] Sixty-one SCOs transferred[37] but for simplicity, I will continue to refer to them as SCOs in this chapter. The SCOs thought that the first problem had also been solved when in 1966, a statement appeared in the *Lancet*: 'The JCC and the Health Department have agreed that it may well be proper for some accident and emergency or casualty departments to be the responsibility of a consultant giving all his time to this work. These appointments must be advertised in the usual way, and senior casualty officers are eligible to apply.'[38] Unfortunately, the Department of Health took no further action and failed to tell Hospital Boards about it,[39] so nothing happened.

Casualty Surgeons Association

The BMA felt that they had achieved what they set out to do and wanted to wind up the SCO Subcommittee. In 1966 David Caro and Edward Abson sought advice from Professor Hedley Atkins (President of the Royal College of Surgeons) who advised them to form an association of casualty officers. The fact that the SCOs were meeting for political purposes had other benefits in that they had started to have clinical meetings. The first in 1966 in Derby was attended by 57 doctors[40] and at the second in March 1967 in Luton, it was agreed to form an association.

The first committee meeting of this Association was held at the British Medical Association on 12 October 1967. Maurice Ellis was elected chairman and a list of attendees is given in the appendix. They unanimously agreed to call themselves the Casualty Surgeons Association (CSA) and to have two meetings a year.

The aims of the CSA were:

1. To promote interest in and by means of investigation to further the knowledge of A&E work in the hospital service.
2. To improve and maintain the standard of work carried out in A&E departments.
3. To protect the interests of those who work in these departments.

At the first meeting they decided to investigate the misuse of casualty departments; the time spent by casualty officers in specific duties and the ultimate source and reason for attendance.

The first Annual General Meeting of the Casualty Surgeons Association was held at Walsall General Hospital on 23 March 1968 at which Maurice Ellis was elected as President, David Caro as Vice President, Edward Abson as Honorary Secretary and John Hindle, Honorary Treasurer.

A&E in the late 1960s and early 1970s

Meanwhile standards in A&E departments were getting worse. The *Lancet* in an editorial described '... dirty, old fashioned, and inconvenient units, staffed by an occasional "consultant-in-charge" but run by victims of the Royal College of Surgeons' insistence on accident work as part of fellowship training, plus ensnared Commonwealth graduates and willing but overstretched nurses.' It stated that the main reasons for the lamentable and worsening position of accident departments were 'patients (too many and of the wrong sort), medical manpower (too little and too poor), accommodation (out of date and dirty), specialist services (inadequate), accident beds (insufficient), supporting staff (too small), administration (often non-existent) and planning (total absence)'.[41]

The staffing of A&E departments in 1969 was shown by a survey conducted by the BOA.[42] The supervision of departments is shown in Table 2.1, which shows that most were in the charge of orthopaedic consultants but only 10 per cent of these departments had paid consultant sessions. In departments with paid orthopaedic sessions, the average was four consultant sessions per department. Table 2.2 shows the number of staff employed and the hours worked. 63.9 per cent of all the casualty officers were SHOs. It also shows that the lower the grade of doctor, the more hours they worked and, in particular, the more out-of-hours work they did.

It was still difficult to find junior staff for casualty departments. The BOA report found that a third of all SHO posts were vacant for a month or more during 1970[42] and on Whit Monday 1971, 13 casualty departments were closed in London alone because of staffing difficulties.[43] In the same year the *BMJ* reported that questions were asked when a medical student was drafted into the casualty department at Birmingham

Table 2.1 Supervision of casualty departments 1969

Specialty	No. of depts	No. of depts without paid sessions	No. of depts with paid sessions	No. of paid sessions
Orthopaedic surgeons	174	156	18	73
Others	38	28	10	66
Joint responsibility	5	4	1	16
No cover	11	11	—	—
Total	228	199	29	155

Source: Reference 42.

Table 2.2 Staffing of 228 A&E departments 1969

Grade	Posts	Av. hrs/week	Av. hrs ENW*	% of total staff	% of total hours	% of total hrs ENW*
HO	50	117	83	6.2	10.0	11.2
SHO	514	73	51	63.9	64.7	70.1
Reg	73	67	41	9.0	9.3	8.1
SR	2	88	0.3	0.3	0.3	0.4
MA	90	52	16	11.2	8.1	3.9
CA	62	46	15	7.7	4.9	2.5
Cons	14	1.7	49	7.0	1.2	0.3
Unpaid HO cover					2.4	3.6

Note: * ENW is evenings, nights and weekend.
Source: Reference 42.

General Hospital when they could not find a locum.[44] 'The difficulty of staffing an accident and emergency department with junior staff below registrar grade is becoming acute.'[45] It was also difficult to fill vacancies with GPs. Fifty per cent of hospitals found no GPs willing to do any sessions with 70 per cent finding no GP willing to do evenings and weekends and 77 per cent not being able to find GPs to work nights.[42]

The unpopularity of A&E posts was not surprising as a working party conducting a 'time and motion' study in 1968 showed that doctors working in A&E got 7 minutes of teaching per week compared to 29 minutes for those in general surgery and 75 minutes for those in general medicine. They spent 1 hr 55 minutes in clinical discussion (compared with 2 hr 24 minutes and 4 hr 10 minutes) and spent 17 minutes per week studying during working hours (compared to 71 and 103 minutes). They did, however, have a shorter working week than those in other specialties. 'A&E jobs were those which nobody wanted and were taken as a last resort. Some doctors admitted that much valuable experience could be gained...but expressed a preference for posts in which accident and emergency work was combined with another specialty.' They felt that they '... lost experience and training because they were isolated from the general work of the hospital in the wards, operating theatre and clinics' and 'they very often had little support from more senior doctors'. 'They were also concerned with the pressure of the medico-legal aspect of their work.'[46]

The reason for these problems was that in most departments there was no consultant supervision. The study mentioned above said that the consultant in charge 'did not appear to the [junior staff] to make a marked impact on the day-to-day work' and 'doctors in almost all Accident and Emergency departments claimed that their consultant hardly ever visited the department.'[46] The *BMJ* agreed that consultant participation was generally inadequate.[26] Scott's survey in 1969[42] (Table 2.1) showed how few funded consultant sessions there were in A&E. In 1970 in a survey of 279 major departments, the Department of Health and Social Security (DHSS) found that less than 80 per cent had a consultant in charge and only 50 per cent had a consultant who worked one session or more in the department. Twenty per cent had no consultant working sessions or on call.[47] A comparison of these two surveys will indicate that many orthopaedic surgeons were working unfunded sessions in A&E. An anecdotal example of the lack of consultant involvement in A&E departments is given by Watts who describes a man who had been an A&E registrar for three months: '... he had occasion to attend a party... where he was introduced to a surgeon on the staff of the hospital who enquired "what are you doing?" to which my friend replied "I am your registrar"'.[48] This type of consultant cover for A&E departments was described by Maurice Ellis as the 'absentee orthopaedic landlord',[49] a description much used and loved by A&E doctors.

The departments were still being kept going by the requirement that candidates for the FRCS should have six months experience in A&E but there was little formal training or teaching. In 1960 the Nuffield Report had recommended 'a closer enquiry by the Royal College of Surgeons about the supervision of designated training posts. If there was a possibility that recognition would be withdrawn if supervision was not satisfactory, the effect could only be salutary.'[5] Nothing happened and in 1970 the Accident Services Review Committee too recommended that 'the Royal College of Surgeons criteria for recognition should be very much strengthened'.[6]

A demonstration of the benefits of a full time consultant over management by committee occurred on the retirement of Maurice Ellis from his post as a full time consultant in Leeds in 1969. In a letter to the *BMJ* he describes how the A&E department was placed under the control of the orthopaedic department rather than any one consultant. Standards deteriorated and, in a department which had never had any difficulty in obtaining junior doctors, no local candidates applied for junior posts. In August 1970 no candidate at all applied but when a new consultant (David Wilson) was appointed, the standard of work improved and local graduates started to apply for jobs.[50] A similar problem was averted at the Glasgow Royal Infirmary where Alec Murray, the consultant in charge was approaching retirement. A history of the hospital records: 'more difficult was to challenge the Scottish Home and Health Department's decision that there was no longer any need to retain the posts of administrative charge of wards'. This was challenged (or ignored) and Mr Murray was replaced.[51]

Despite this it must be remembered that there were some good A&E departments run as part of accident services. In 1973 the President of the BOA stated: 'Such large accident centres can, we believe, be accepted as running a service of which we can be proud...Morale is high, policy agreed and supervision and involvement of the consultant surgeon play an essential part in the successful operation.'[52] The Department of Health also recognised that there were some departments where orthopaedic surgeons (and occasionally others) have shown a very great interest in A&E and had created centres of excellence.[47]

Lack of consultant cover in most A&E departments was not entirely the fault of orthopaedic surgeons. The lack of funded sessions has been noted and 50 per cent of hospitals were below establishment for orthopaedic surgeons[29] at a time when orthopaedic workload was increasing with more operative management of fractures and joint replacement surgery. However orthopaedic surgeons, not unnaturally, preferred orthopaedics and admitted it: 'inadequacy of consultant supervision is due in part to inadequate numbers in the specialties and in part due to a lack of interest on the part of individuals and organisations...An orthopaedic surgeon working "half time" in the hurly-burly of a Victorian casualty department poorly converted into an accident service and "half-time" in the relative peace of a long stay orthopaedic hospital or ward, and finding that each "half" is almost a full-time job, may provide inadequate supervision.'[53] Other surgeons wrote: 'an orthopaedic or general surgeon having facilities for the pursuit of his specialty is likely to forgo the boredom of the predominating trivialities of casualty...nothing could be more calculated to break his spirit than to have to give a large measure of his time to routine casualty sessions.'[54] The official history of the BOA states: 'it is probably fair to suggest that as a whole, orthopaedic surgeons were unwilling or unable because of other commitments, to devote sufficient time to what became the lesser of the two component parts of their specialty'.[55]

It was recognised that the situation was desperate. The official BOA view was that 'in general the need for consultant service within these Casualty Departments is best served by the appointment of further orthopaedic surgeons'.[52] However most orthopaedic surgeons did not want the responsibility for dealing with non traumatic emergencies: 'The junior staff of the accident and orthopaedic department who man the accident unit cannot cope with both traumatic and non-traumatic emergencies, particularly if the latter require detailed and time-consuming examination.'[56]

One solution was a split service with an accident centre treating trauma and other problems being treated elsewhere.[57] Orthopaedic surgeons felt that general physicians and surgeons were shirking their responsibilities. At a symposium on A&E staffing it was said: 'it is high time that our general surgical and general medical colleagues took responsibility for the reception of the emergency problems relevant to their specialties. Moreover, it is time that they contributed to staffing, at all levels, of the reception areas for their own particular emergencies,'[31] and in summing up the conclusions of the symposium, O'Connor (an orthopaedic surgeon) proposed that a hospital should have no fewer than four emergency departments. These would be for trauma, general medicine (including psychiatry), general surgery (including gynaecology and obstetrics), each staffed by the relevant specialist firms and a fourth one for general practice type cases. This would be staffed by GPs and physicians or by residents of all specialties on mandatory rotation under the direction of a hospital GP or a physician. Thus casualty officers could be abolished forever.[31] Da Costa had previously noted that this approach would be very costly in manpower and while it might work in teaching hospitals, it would not work in smaller units[58] and Caro had pointed out that few hospitals could justify even two resuscitation rooms.[59] The Todd Report in 1968 had also called for a single arrangement for managing emergencies: '... we do not believe that the present requirement for the most junior members of each small clinical unit to be on call every night can be maintained much longer; ... economy of space, equipment and personnel is likely to lead to more widespread establishment of single instead of multiple arrangements for the reception and management of emergencies of all kinds.'[60]

A second solution put forward largely by orthopaedic surgeons, was for A&E departments to be run by GPs, probably on a sessional basis: 'General practitioners should be invited to staff casualty departments'[61] and 'our casualty departments need high-grade "general duty medical officers" as daily supervisors of the work and teaching. I am sure of this need as long as men are not divorced from work of domiciliary type and from good health centres.' However, overall charge would still be in the hands of a consultant.[62] Advantages of appointing GPs to this job with no job satisfaction were that 'a sturdy foundation could be built without the need to appoint a consultant to a job which rarely calls for such a level of experience'[63] and that 'a doctor can expect to pass beyond such excessive commitments after a few years service.'[64] The only problem with this as a solution is that nobody had asked the GPs and they did not want to provide 24-hour cover to A&E departments.[65] Not all orthopaedic surgeons were against A&E as a specialty. Charnley recommended in a letter to the *BMJ* that 'casualty departments (and first attendance accident services) must be staffed by men making it a permanent career'[66] but made no comment on the grade to which such doctors should be appointed.

The BOA recommended that an experienced doctor should be on reception duty at all times and this required a minimum of four doctors to provide 24-hour cover including annual leave, study leave and sickness (if a 40-hour week was accepted, five doctors would be required). These doctors should be recruited from general practice and have pay and conditions similar to those in general practice. Noting the fact that more senior doctors tended to work fewer evening, night and weekend hours, they recommended that to ensure 'harmonious running of the department unattractive shifts should be shared equally between all staff who must therefore be of equal status'. They warned against the 'establishment of a privileged weekday worker, who may deter recruitment of the other men required to man the department for the remaining 128 hours of the week'. These doctors could supervise SHOs and even PRHOs. There would be a consultant orthopaedic surgeon in charge of whole accident service but flexibility was required in different parts of the country.[42,67]

Many orthopaedic surgeons were still vehemently opposed to A&E consultants. Perceived potential problems with appointing such consultants were that it 'sacrifices the principle of planned continuous care of the patient by a single clinical team ...' and might lead to a deterioration in the treatment of the injured. It 'may increase the misuse of casualty departments' and an A&E consultant might operate a 'consultative, minor operating, and follow up clinic for non-emergencies in his own field of interest; to the detriment of the prompt and efficient treatment of the injured ... '.[42] In addition 'four or five co-equal consultants would not appear to be a reasonable alternative' to four or five subconsultants working under an orthopaedic surgeon.[52]

Others described the type of appointment recommended by the BOA as a 'hospital GP'. 'It will be the source of demoralization and confusion to attempt to use the status of consultancy to bait men to work where neither pay nor conditions will do so. Why not call him [the hospital casualty officer] a hospital GP?'[64] Another felt that this person needed consultant pay without specialist connotation and could be called a 'senior emergency officer'.[68]

However by about 1970, there was growing pressure for consultants in A&E. A significant factor was the constant pressure by a small group of SCOs. Among the more vociferous were Abson and Caro who, apart from their political activities described above, argued and lobbied for the creation of consultants in casualty in an article in the *Lancet*,[69] at symposia and in letters to the medical press.

It was generally accepted that after eight years, the Platt Report had been given a fair trial and that orthopaedic surgeons had 'failed to develop a trouble-free casualty service, and while this may not be entirely their fault, it is their responsibility because they accepted this burden'.[70] The *Lancet* asked: 'Is it not time to acknowledge that this important aspect of hospital practice also deserves its own full-time consultant cover?' and said '... when the emergency services are in the hands of experienced and interested doctors, the problems...will take care of themselves'.[71] It later said: 'the crucial first step in reform must be to ensure that every department has at its head a consultant whose duties lie primarily or entirely within it'.[41] The *BMJ* said: 'One possible solution of ending the second class status might be to have a career grade supporting the consultant, with the opportunity for suitably experienced staff to achieve consultant status in the accident and emergency department'[26] and later: 'the accident department requires a consultant to take full charge of it to give the department equality with other specialties'.[43]

The Department of Health was also concerned and in 1968 said: 'the personal day to day supervision of the work of the department by its consultants still needs to be emphasised'[72] though the *Lancet* pointed out: 'the injunction to regional hospital boards to ensure that the personal day-to-day supervision of the accident and emergency department is undertaken by its consultants is a little curious. These departments have no consultants in the general sense of the term.'[71]

The problem was what kind of person this consultant would be and possibilities were discussed in articles and letters to journals. One solution was a traumatologist. 'He need not displace the orthopaedic surgeon, the neurosurgeon, anaesthetist or any other specialist....He would be a much-needed though new type of general surgeon who would still require their real expertise.'[73] To answer the problem that many of the patients were not trauma patients, it was argued that 'a good man, well trained to deal with accidental injuries, will usually be able to cope adequately with most emergencies ...' but he shouldn't normally be expected to deal with cases which properly belong to other specialists.[73] Even the *BMJ* wrote: 'the more satisfactory solution would be to encourage the suitably qualified to develop an interest in the management of trauma within the accident department and to recognise this with consultant status.'[43] Some SCOs also favoured traumatology as a separate specialty.[74]

A second possibility was to appoint a junior surgeon who would run the A&E department for a few years before moving on to a post in one of the branches of surgery[75,76] but this would have perpetuated the idea of A&E work as being less important.[43]

A third view was that A&E consultants should be GP-type doctors and one correspondent to the *BMJ* thought that no special training was required 'beyond that learnt as an undergraduate and as a house surgeon, except perhaps some advanced first-aid techniques such as the confident passage of an endotracheal tube'.[77]

Despite these various views, it is possible to see from the advice being proffered at the time, the origins of the A&E consultants who were later appointed. Garden felt that it was unimportant whether a consultant was a general or specialist surgeon. The main essentials were that he should be interested and hold himself in constant readiness to meet the appeals of his junior staff.[34] He would be 'a physician, surgeon and psychologist all in one, prepared to allay the anxieties of the patient suffering from a trivial complaint with the same assurance as he deals with barbiturate poisoning, cardiac arrest, respiratory obstruction, or haemorrhagic shock'.[78] The SCOs in 1963 at their first meeting with the CCSC had stressed that consultant posts 'should be given to people with broad, rather than specialised training and experience'.[19]

The Lancet believed that 'the ideal casualty officer...is neither physician nor surgeon but a combination of both. Working in a service which leans heavily on the good will of every other hospital department, he requires an abundance of tact and understanding: and much rests on his ability to cooperate with his inpatient colleagues on whom the ultimate success of his efforts in early diagnosis and resuscitation depends.'[71]

Others also stressed the importance of training in resuscitation[79,80] and this also appeared to be the view of the Department of Health.[47] In the light of their emphasis on the importance of the 'emergency situation', it is not surprising that the SCOs themselves thought that their specialty should be based around resuscitation.[74] One senior orthopaedic surgeon thought that far from being an expert in resuscitation, 'the casualty consultant would the expert in dealing with the less serious conditions'.[81]

Other important functions were organisation and teaching. The *Lancet* felt that 'his main function would be to teach and organize a service. His own training would have equipped him to communicate with colleagues with suitable authority. ... His main duties would consist of day-to-day supervision of the work of successive generations of senior house officers preparing for general practice.' He would also be involved with pre-Fellowship teaching and teaching undergraduates, nurses and ambulance personnel.[70] Indeed the ability to teach and train should be one of the major criteria for appointment.[70] It was recognised that the A&E consultants by themselves would not meet the service needs of the departments but that by improving standards, posts would become more attractive to junior doctors[43,70] and the costs of employing consultants could be balanced by 'savings to be gained from a reduction in morbidity, reduction in medico-legal claims, and a reduction in the number of unnecessary Xrays'.[43] It was recognised that A&E consultants might develop a special interest in hand or head injuries.[79] A higher qualification was thought desirable[47] or essential.[80]

Surprisingly in 1971 the BOA (possibly aware that A&E consultants were inevitable) also unanimously supported the idea but their concept of A&E consultants had very little in common with the usual meaning of the term. They stated: 'We do not support the creation of casualty consultants, we do support the view that there should be a consultant in accident and emergency medicine but he should be in charge of the whole accident service including in-patient beds and out-patient clinics, a function largely taken on by orthopaedic surgeons. ... The staff of the primary care department or reception unit of the accident and emergency department should be under the control of these consultants and must therefore be of subconsultant grade.'[29]

The Platt Report had made much of the fact that 61 per cent of the work of an A&E department was trauma but the medical workload was increasing and medical patients took more time to deal with than trauma patients and tended to cause more difficulties. If one reanalyses the admissions detailed in the Platt report, the workload is more evenly matched with 35.2 per cent of admissions due to trauma (only 24.1 per cent due to fractures, dislocations and multiple injury), 29.4 per cent medical and 25.7 per cent surgical non-trauma. In London the proportion of medical patients was higher. The Royal College of Physicians (RCP) appears not to have been involved in discussions about A&E departments in the 1950s or for most of the 1960s. However in early 1969 they set up a committee under the chairmanship of the president, Sir Max Rosenheim, to look at medical work in A&E departments and this reported later the same year. They reported that in London more than half of admissions from casualty were medical or paediatric.[82] (It must be remembered that at that time the Royal College of Physicians was a home for paediatricians as well as physicians.) They noted that there were problems with the management of medical emergencies in casualty. This was because of a surgical bias in junior A&E staff, poor supervision by senior medical staff and the fact that physicians on call for medical emergencies often had other commitments and might even be at another hospital. They recommended that a consultant physician should take responsibility for arrangements for the management of medical (including paediatric and psychiatric) patients; supervising clinical management; drawing up 'medical' standing orders; ensuring smooth transfer of medical patients to an appropriate ward and teaching. They also recommended that where the volume and nature of medical caseload justified it, the A&E establishment should include one member of staff who was medically orientated. This could be a medical registrar or SHO on rotation or a doctor in a career grade (they almost certainly meant a Medical Assistant). The importance of medical involvement in A&E departments was further emphasised by a study of patients passing through a resuscitation room and published in 1972. In 129 days, 216 medical, 142 trauma and 16 surgical and gynaecology patients were seen with a higher mortality rate in medical patients.[83]

In July 1969 the Representative Body of the BMA (prompted by much pressure from the senior casualty officers) passed a resolution to the effect that the action being taken to secure a career structure in A&E work rising to consultant status for major A&E centres was grossly inadequate. A working party of the Central Committee for Hospital Medical Services (of the BMA) (CCHMS) was set up which met seven times and reported in 1971. This had more SCOs than orthopaedic surgeons. Two of the conclusions were: 'Consultant posts (including those carrying administrative charge) should be established in accident and emergency departments ...' and 'a training programme for accident and emergency work should be provided ...'[84] The CCHMS was almost equally divided between those who felt consultant posts should be established and those who felt that there was no place for consultants. However it agreed that a pilot scheme should be set up to appoint consultants where this could be done without detriment to the standard of consultant appointments generally. It recommended that sessions should be given for teaching. Hospitals with A&E consultants should be monitored over 3–5 years and compared to other departments.[85] If this monitoring showed that A&E consultants were viable, then a formal training programme would be necessary.[86]

In January 1971 the JCC received this report. It said that it was opposed to doctors of any specialty being in nominal charge but 'that there is a case for examining, on an individual and ad hoc basis, the position of some senior and experienced casualty officers devoting virtually the whole of their time to the service but not of consultant status'. Also 'that further consideration be given to the recognition of A&E Medicine as a specialty in its own right in an endeavour to find basis for agreement, and to the related problem of the training programme that would be appropriate for such a specialty; and that a special subcommittee be appointed for the purpose by the Chairman'.[87] This subcommittee first met in March 1971 under the chairmanship of Sir John Richardson. There were two representatives of the CSA, Maurice Ellis and Alec Murray. This recommended that some posts of consultant in A&E be established and monitored over three to five years as a pilot scheme. A higher qualification was desirable but not essential and there should be sessions for teaching. Consideration should be given to the integration of non-surgical specialties to ensure that consultants in all disciplines accepted a responsibility to back up the A&E department. It also felt that there was no need to consider a training programme at that stage until the pilot schemes had been evaluated but preliminary consideration to training schemes should be given by the Royal Colleges.[88] The idea was also supported by the Council of the Royal College of Surgeons of England.[89] Following further meetings of the subcommittee to work out details, the proposal to appoint 32 consultants were given the go-ahead at the end of 1971. This is often called the Bruce Report.

The DHSS and profession agreed to appoint 30 posts initially. Applicants would need a breadth of experience of A&E work and a higher qualification was desirable. It was important that standards should not be lowered.[47]

Letters were sent out by the Chief Medical Officer to the Senior Administrative Medical Officers of the Regional Hospital Boards inviting them to apply for these posts. I am uncertain as to how this happened. The letter to the South West Regional Board was sent on 12 November 1971 but by this stage a consultant in Walsall (Michael Merlin) had already been appointed. All posts were to be advertised and the job description 'should specify responsibility for the primary care of emergency cases. The posts...should entail the consultant normally spending the bulk of his time in the emergency department'. A higher qualification was desirable but not essential. The letter also contains an interesting paragraph: 'It is proposed that the posts will be defined and advertised as for consultants in charge of emergency departments, without a specialty being specified. The question whether emergency work constitutes, or should constitute, a separate specialty involves consideration by a number of professional bodies of such matters as training programmes. The decision to ask for proposals from Boards cannot prejudge the outcome of that discussion, though experience of departments in which the new posts are created is expected to be a useful contribution to it.'[90]

3
The First Consultants

As has been noted, the first full time consultant in casualty was Maurice Ellis at Leeds General Infirmary. He was placed in charge of the casualty department in 1949 while he was a general surgical senior registrar and was appointed consultant in casualty in 1952. David Wilson, later a president of the Casualty Surgeons Association (CSA), was a medical student in casualty in Leeds when Mr Ellis was first appointed and says that the department was transformed by his arrival.[1] On his first day Mr Ellis insisted on cleanliness with old dressings being removed from the floor and he created organisation where there had been chaos. A description of the organisation of his dressing clinics is given in a paper he wrote on hand injuries.[2] He also introduced audit and research on hand injuries,[2] antibiotic use[3] and tenosynovitis.[4] He retired in 1969 and following the Platt recommendations, was replaced by an orthopaedic surgeon with the department being managed by the orthopaedic service rather than by an individual consultant. He, himself, in a letter to the *BMJ* describes how standards fell and the hospital realised their error and appointed David Wilson, by then a locum senior registrar in orthopaedics, to the consultant post the following year following which standards started to improve.[5] Shortly before this, in 1969, Howard Baderman had been appointed as an A&E consultant at University College Hospital, London (UCH) when a new department was opened. He had previously spent eight years as RMO in the hospital where one of his responsibilities had been to be in charge of A&E and he had advised on the design of the new department.[6]

Prior to this it is difficult to define an A&E consultant as there were a few surgeons who devoted a substantial part of their time to casualty (e.g. Patrick Clarkson and T.G. Lowden – discussed previously). Various documents report that prior to 1970 there were three or four consultants in casualty. In addition to Maurice Ellis, I believe that these refer to Alec Murray at Glasgow Royal Infirmary and David Proctor in Aberdeen. Alec Murray was appointed as a surgeon with responsibility for casualty in 1960[7] and appears to have come to spend all his time in the casualty department and thus became a full time casualty consultant even though that was not the post to which he was appointed. Mr Proctor's obituary[8] says that he was a senior casualty officer and later consultant in A&E care 1952–81. The SCO grade was not started until 1953 and he must have been an SHMO who was upgraded to a consultant (probably without an interview) in a way that could not happen for SCOs. The date of this upgrading is uncertain but his successor thinks that this was probably before 1965.[9]

The main breakthrough came in 1971/72 when 32 consultants were appointed as a pilot study with plans that departments with consultants should be monitored over 3–5 years and compared to those without consultants.[10] They did not just have to convince the Department of Health but they also needed to convince their colleagues in other specialties and, especially, orthopaedic surgeons who had predicted disaster.

The opposition of the British Orthopaedic Association to the idea of A&E consultants has been noted and they also advised: 'The appointment of selected "casualty officers" as consultants in the accident and emergency departments should be undertaken with caution' as they considered that there were insufficient suitably qualified people to fill the posts.[11,12] Another surgeon predicted in an article that 'advertisements, if they appeared now in the medical press would bring applications from totally unsuitable individuals'.[13]

Michael Merlin, probably the first of the new consultants was appointed in Walsall in 1971 and most of the rest during 1972. There were 22–68 applicants for each post, though numbers of applicants fell with time.[14] The Lewin Report[15] (described below) gives details of the appointees. Of 32 posts advertised, two appointments were not made initially. Sixteen posts were filled by Medical Assistants at the hospital where they were already in post (e.g. John Collins in Derby, Malcolm Hall in Preston). Eight more were filled by MAs who moved (e.g. David Caro, St Bartholomew's and Edward Abson, Canterbury). For hospitals seeking to improve their staffing, there was an incentive not to appoint an MA who was already in post at the hospital as the letter advising that appointments could be made stated that if 'the Medical Assistant obtains the post, his own post should not normally be refilled on a permanent basis without consulting the Department about other possibilities; the post might not need to be filled, or a Senior Registrar post might be appropriate'.[16] Thus 24 of the 30 posts were filled by experienced doctors already working in A&E. Not all appointees had this experience: two had a largely general practice background. Ian Stewart (Plymouth) had a number of years of experience as a surgical and orthopaedic registrar and had taken a post at the Birmingham Accident Hospital in 1971 to prepare for the new posts which he knew would be advertised.[17] Three were women. Twelve had an FRCS, two were Member of the Royal College of Physicians (MRCP), one had a Diploma in Public Health and one an MSc in Occupational Medicine.

David Caro's obituary states that his appointment initially met with 'entrenched opposition'[18] and he was almost certainly not unique in this. Not surprisingly, in view of the opposition to A&E as a specialty, the first consultants had little status. Dr Gerard Vaughan, then MP for Reading, giving evidence to the House of Commons Expenditure Committee was quoted in the *Lancet* as saying: 'Here is the front line, or one of the front lines, of the hospital. It is also one of the most difficult parts of the hospital, it is also one of the most urgent parts of the hospital, and yet into it go the least trained, and it is the least popular department. I wonder, really, if medically we have not gone very wrong somewhere on the educational side within our profession. The casualty surgeon, for example, although he is a consultant, is a dogsbody, is he not? He has the least status of all the surgeons in the hospital.' The Royal College of Physicians agreed.[19]

The newly appointed single-handed consultant also had other concerns such as who would cover him when he went on annual or study leave[20] or even who would cover him for a night or weekend. Ian Stewart was initially on call for 12 nights out of 14.[17]

Correspondence to the *Lancet* showed that the appointment of consultants without a higher qualification upset some: 'Is the acquisition of consultant status without postgraduate letters-patent going to be extended to other unattractive disciplines? The profession should think on these things.'[21] Others defended the decision to appoint consultants without a higher qualification.[22]

Almost as soon as the first 30 consultants were in post, other hospitals started looking for A&E consultants. Dr Baderman reported at a CSA committee meeting in October 1972: 'It was felt that now the first wave of consultant appointments was almost finished, inevitably a second wave would follow due to the stimulated enthusiasm of those departments which at the moment had not been able or had not attempted to get a consultant.'[20] It was too early by this stage to claim that this was evidence that the experiment had already proved successful and was more likely to be due to the hope that a consultant would solve a department's problems. However the benefits of A&E consultants soon did become apparent. The DHSS reviewed the pilot sites in 1974 and reported that departments with consultants were better at attracting junior staff and concluded: 'in no instance has an appointment failed to achieve some positive benefit. In a number of instances there have been significant improvements in the organization of accident and emergency services in its wider connotation.'[14] They did however find that most had been appointed 'without being given a clearly defined brief relating to their responsibility for organising accident and emergency services'. One consultant described to me his job description as 'a blank sheet of paper'.[23]

The profession as a whole soon accepted this new specialty. Following the Lewin Report in 1978, the CCHMS sent a questionnaire to its Regional Subcommittees and reported that: 'There was a unanimous view that accident and emergency departments should have effective consultant supervision provided by consultants in accident and emergency medicine.'[24] The *BMJ* wrote: 'The creation of consultant posts ... has improved the status of the work and helped to upgrade many departments. Health authorities and now even hospitals themselves – despite long traditions of vested interest in casualty on the part of some specialties – have begun to ask for accident and emergency consultants.'[25]

The number of posts therefore continued to grow as shown in Table 3.1. Most consultants had a surgical training. This was for a variety of reasons. A&E was still seen as a surgical specialty and was frequently listed as a surgical subspecialty until the mid-1980s. All surgeons would have had basic A&E experience at SHO level as a requirement for sitting the FRCS exam whereas most physicians would not have. In addition the number of openings outside surgery for those with

Table 3.1 Number of A&E consultants

Year	Number
1973	57
1974	80
1975	91
1976	105

Source: Lewin Report.[15]

an FRCS was very limited whereas the MRCP was useful for careers in general practice, occupational medicine and many other specialties. There was frequently a misapprehension by other specialties that registrars in surgical specialties were ideally suited to be consultants in A&E[26] and many who had failed to progress in their career tried to transfer to this new specialty. Department of Health statistics show that in 1976 there were 24.7 applicants for every A&E consultant post compared to 5.5 for all specialties combined. Most candidates apply for more than one post and there were 7 candidates per post (2.3 for all specialties combined). The average age of appointment to an A&E post was 42 years reflecting that most had done something else first (37.2 for all specialties combined).[27] Similar figures applied for 1977 and 1978.

A number of the early A&E consultants had experience in the missions (or similar overseas experience) including David Wilson and Michael Flowers (Leeds), William Rutherford (Belfast) and Chris Cutting (Taunton). By going abroad they had fallen off the conveyor belt of teaching hospital posts leading to a senior registrar appointment in a mainstream specialty but also their very wide, general experience was valued for A&E. Also considered of value was military experience.

If one makes appointments from a relatively static pool of individuals, it is almost inevitable that the best qualified will be appointed first and that subsequent appointees will be less well qualified. In A&E the first 30 appointments were probably all of people well qualified either by experience or training, but the supply of well-qualified individuals was being used up with no one being trained. The president of the BOA wrote as early as 1973: 'With some of the holders of the new emergency consultant posts – lacking all distinction from higher qualification, special training or even special experience – the distinction between the consultant and the hospital practitioner grade doctor within the same department may be difficult to explain and justify.'[28] The longstanding opposition of the BOA to A&E consultants meant that this could be dismissed as political posturing but soon others were saying similar things. In early 1975 a working party looking at setting up senior registrar training, warned 'if standards in this specialty are to be maintained it is strongly urged that the number of consultant appointments made should be restricted as it is clear that the number of those likely to apply with the necessary qualifications and experience is now limited. This restriction should apply until there has been time to implement a training programme as now recommended.'[29] The *BMJ* stated that some appointments were said to cause concern[30,31] and Peter London, a respected surgeon from the Birmingham Accident Hospital, said: 'For many of those appointed to accident and emergency posts the work involved has neither been their first choice nor that for which they have been trained, and the significance of the fact that many advertisements have stated that a higher qualification is not essential is obvious and not encouraging.'[32] Prof. James, the chairman of the Specialist Advisory Committee (SAC) in A&E Medicine (and a supporter of the specialty) wrote: 'Many consultants in this new discipline have been appointed in the last few years, some with inappropriate training because there has been no established pattern. As a result of this and even more because of an initial failure to define the function of this specialty there have been many misunderstandings and difficulties following recent appointments.'[33] At a CSA committee meeting Mr Wilson 'had to agree that some unfortunate appointments had been made but that this was in part due to the great demand for A&E consultants'.[34] Not surprisingly, consultants are not prepared to identify which of their colleagues were appointed inappropriately nor those who did the job inadequately and I have no wish to be sued for libel! However this was clearly perceived to be a major problem and it is important to discuss it.

London's concern that A&E was not the first choice of career for most appointees was certainly true but is surely unfair as until 1972 no consultant posts in A&E existed for people to aspire to. Many doctors change specialty for one reason or another and devote themselves to their new career and David Caro pointed out to orthopaedic surgeons at a symposium, that at that time many orthopaedic surgeons probably started out intending to do general surgery.[35]

It is also true that many did not have a higher qualification. At that time there was no specific higher qualification for A&E and either FRCS, MRCP or FFARCS would have been sufficient to satisfy the critics. (Membership of the Royal College of General Practitioners (MRCGP) was probably equally appropriate but would not have satisfied many hospital doctors.) It was therefore not possible to say that the absence of a qualification meant that an individual lacked the knowledge to become a consultant in A&E. A higher qualification of some sort was therefore just an intellectual hurdle to be jumped. Its lack did not make a consultant less able to do his job but, combined with the fact that the A&E consultant had not done the SR post which consultants in other specialties had done, it devalued him in the sight of his colleagues. This might therefore weaken A&E's negotiating position within the hospital and could be seen as inappropriate.

Prof. James's concern that consultants were appointed 'with inappropriate training because there has been no established pattern' was due to delays in instituting a training programme. He also identified misunderstandings in the role of the A&E consultant. Chapter 2 noted confusion over the meaning of 'accident and emergency services'; the view of a number of people that the A&E consultant should be a traumatologist and the BOA's view in 1971 that the consultant in A&E should be a consultant orthopaedic surgeon. It is quite possible that some hospitals were either confused as to what post they were appointing to or were deliberately misled into appointing someone to reinforce orthopaedic numbers. One job description read: 'It is envisaged that the routine of the Accident Consultant will be arranged in such a way that he will be able to play a full part in the Accident and Emergency Service, both within the Accident Department and the Wards and Outpatients, with access to beds and to routine theatre sessions, though not to so great an extent as the other surgeons.'[36] There had, in places, been confusion between advertisements for A&E consultants and those for orthopaedic surgeons[37] and there had been concern expressed at the CSA Committee about orthopaedic bias on appointment advisory committees.[20] There would have been very few A&E consultants on appointment advisory committees. Many of the new consultants had been preparing themselves for a career as a 'trauma surgeon' to work in units like that at the Birmingham Accident Hospital. When it became clear that trauma surgery was not going to become a specialty, many moved into A&E. Most probably fully embraced their new specialty with medical emergencies as well as trauma but it would not be surprising if some concentrated on trauma to the exclusion of other aspects of their job.

Early consultants who had trained in another specialty before moving to A&E, brought with them those skills, and sometimes these matched needs within the hospital where they were appointed. Thus David Wilson and Michael Flowers in Leeds performed hand surgery and Ian Stewart in Plymouth looked after abdominal trauma. Many A&E consultants looked after head injuries and a number did some orthopaedic surgery. The Mills Report noted: 'We are aware that in some of the smaller departments ... many Accident and Emergency consultants undertake a wider range of treatment particularly in the field of orthopaedic surgery ... .'[38] It recommended that in smaller departments consultants could have '... a firm but limited number of sessions in the appropriate specialty in which the candidate has previous experience'. However this should not 'undermine the essential principal that a consultant in an Accident and Emergency department should spend the majority of his time actually working in the department and we would consider that a minimum of 6 sessions is necessary to fulfil this fundamental duty'. This appears to be recommending that hospitals should appoint an A&E consultant and then find appropriate other sessions for him to do but some hospitals continued to advertise for an A&E consultant to join the orthopaedic rota certainly as late as 1983.[39]

Working within another specialty was acceptable as long as it could either be done within the A&E department or took a relatively small proportion of the working week as otherwise the A&E consultant, himself, could become an 'absentee landlord'. A few individuals started to practise mainly in their original specialty. Appointing a doctor who would never have obtained a consultant post in orthopaedics to a consultancy in A&E from which he spent a substantial proportion of his time doing orthopaedics could fairly be described as inappropriate.

Another complaint was of inconsistency by appointments committees with some appointing MAs with many years clinical experience but without higher qualifications and others insisting on a higher qualification and giving more weight to this than to the applicant's clinical experience.[40]

The senior registrars were particularly concerned about the standard of consultant appointments. There were a small number of senior registrars and a large number of consultants to be appointed and they wondered where the consultants were going to come from. William Morgan, one of the early senior registrars wrote to the *BMJ*: 'My fear is that they will be made up by appointing registrars and senior registrars who, for one reason or another, have not made the grade in their own specialties. If this occurs it will be a disaster for the accident service in Great Britain. ... Just as one would not consider appointing a consultant in general surgery or orthopaedics who had not served an adequate time as a senior registrar, so the future accident and emergency consultant should be required to complete an adequate senior registrar training.'[41] There is no doubt that this happened but many of the senior registrars themselves, were only doing a few months of their training programme before obtaining consultant posts and this, too, was seen as inappropriate.

Orthopaedic response to A&E consultants

The BOA had been politically out-manoeuvred in the battle to get consultants in A&E. Orthopaedic surgeons had been outnumbered by SCOs on the CCHMS working party with only a single orthopaedic surgeon representing the CCHMS with no BOA representative and there were only two representatives at the meeting between the JCC and

DHSS. They appear to have resented this in the same way as the SCOs resented their not being consulted by Platt. The BOA wrote: 'It is regrettable that the group of surgeons – orthopaedic surgeons – who in the last decade have borne the brunt of the difficult problem of running Casualty Departments, have been so little involved in the discussions which have led to the experimental creation of this new consultant group.[28] They continued to believe that 'in general the need for consultant service within these casualty Departments is best served by the appointment of further orthopaedic surgeons', but agreed that an 'emergency consultant' might be appropriate in urban areas with a high medical workload. They also admitted 'to misgivings about some of these appointments as emergency consultants, more particularly lest these may interfere with the planned continued care of the injured. ...'[28] The official history of the BOA stated that eventually they grudgingly accepted the need for A&E consultants but wanted to ensure that those appointed would not interfere with the work of the orthopaedic surgeons and so tried to insist on physicians rather than surgeons being appointed.[42]

Individual groups of orthopaedic surgeons reacted in different ways. As noted, some appointed A&E consultants who had orthopaedic training to assist with their workload but others stuck to their principles and insisted on appointing non-surgeons so that they would not interfere with trauma management with the result that at least one person was appointed with no A&E experience and had to be sent for training before taking up post to the resentment of the Medical Assistant who applied for the post and failed to get it. Other orthopaedic surgeons continued to oppose the new specialty: 'Orthopaedic surgeons ... should insist that the management of locomotor injuries *from the outset* should remain their responsibility. ... We feel in Stoke that there is no place in our unit for this post [A&E consultant].'[43] In Nottingham, orthopaedic surgeons continued to do the initial treatment of all fractures despite the early appointment of an A&E consultant and this arrangement persists.[44]

Eventually A&E was accepted but this was gradual. After a meeting with orthopaedic surgeons in Oswestry in 1974, it was reported to a CSA committee meeting that although orthopaedic surgeons were still not sold on the idea of non-orthopaedic surgeons in charge of an A&E department 'the measure of invective was more benign!'[37] Also in 1974 the BOA circulated a questionnaire to orthopaedic surgeons about the A&E consultant appointments in their hospitals. The response rate was poor but most reported that the system was working well though there were some problems with one A&E consultant considering himself a 'traumatic surgeon' and wanting to undertake the full care of injured patients.[42] The war could be said to be over in 1978 when Prof James in his presidential address to the BOA said: '... I have come to believe that the accident and emergency department should in general not be the province of the orthopaedic surgeon and that we should leave this and return to what we are trained to do. ... I am sure that there is now a clearly defined specialty of emergency medicine working in cooperation, not competition, with the orthopaedic surgeon. ... As men become available from the newly approved training schemes in this specialty, and it will take time, I believe we should gracefully withdraw from an area where we have not shown distinction.'[45]

Lewin Report

It had been assumed that the evaluation of the pilot posts would be done jointly between the Department of Health and the profession. Unfortunately the evaluation was done by the DHSS alone in 1974 less than two years after the initial appointments. Even worse, this evaluation did not seem to have been adequately communicated to the profession and 'for reasons unknown, the JCC was unaware of it'.[38] This led to statements such as: 'The stated intention was that 30–40 should be appointed and the results studied as part of a pilot scheme; in fact more than 150 have been appointed with no evidence that any sort of evaluation has been undertaken.'[32]

On 5 July 1977, the JCC discussed the problem of inappropriate appointments and declared a moratorium on new consultant appointments. It set up a working party to look at the problem, though in October, because of the urgency, Walpole Lewin, a neurosurgeon was asked to undertake a one-man enquiry.[15] His report was published the following year with the DHSS evaluation of the original 32 posts as an appendix. He found 'there is overwhelming support for consultant cover in A&E departments to be a reality however it is provided ...' and he also found general support for consultants in A&E 'even among some who had been sceptical at the inception of this grade'. There were however two important provisos. 'In the future consultants must be properly trained for the job' and 'there should be no drive to impose this arrangement as the sole method of running the service.' There was further statistical information on consultants but this adds little to the information found by the DHSS in 1974: about 60 per cent of the, by now 109, consultants had a higher qualification and the vast majority of higher qualifications were FRCS. From this he felt that the majority of A&E consultants would be surgeons but that 'the potentialities of a career in this specialty should also be brought to the attention of physicians and anaesthetists in training'.

He made recommendations on the criteria for appointment. He agreed that higher qualifications were important, if only for the consultant to be seen as the equal to his colleagues in other specialties. Although higher qualifications should not be mandatory, Lewin thought that 'only rarely should an appointment be offered to a candidate who has not a higher qualification of this country'. Experience of A&E work was of vital importance but he recommended that deficiencies could be made up with a proleptic appointment to obtain further experience before taking up the post. He also stressed the importance of administrative and teaching abilities and the personality 'to enable him to liaise effectively and harmoniously with the other departments of the hospital'. He advised that hospitals should have A&E subcommittees to support A&E consultants.

He recommended that the moratorium should continue with 11 new consultant posts in 1978/79 and none in 1979/80 (though retiring consultants and MAs could be replaced by consultants).

The Royal College of Surgeons recommended (presumably to its assessors on Appointment Advisory Committees (AAC)):

> It may be that to require full accreditation in an existing specialty as indicated by the JCC would be excessive even as an interim requirement but you should look for:
> 1. the possession of a higher diploma and
> 2. at least three years of experience relevant to A&E work. In exceptional circumstances one or other of these requirements may be waived, but not both.[33]

I believe that the DHSS had made a decision in 1971 that they were going to appoint A&E consultants and that, as with subsequent political restructurings of the NHS, they were impatient with the idea of a pilot study. In 1971 they were anxious 'not to overstress the experimental aspects' of the pilot study as they believed that it was important for trainees working under the new consultants to see career prospects in the specialty.[46] It is interesting to ask why they did not involve the profession in their 1974 evaluation. In 1978 they were unhappy about the moratorium as they were under great pressure to appoint consultants to improve the staffing of A&E departments. There was also a moratorium on MA posts and so the only way that staffing levels could be improved was to appoint SHOs. The Chief Medical Officer felt that it was inappropriate to say that there were not enough suitable applicants and that one could only find out whether there were adequately qualified people to take up posts by advertising and assessing the applications at an appointments committee though he agreed that if the standard of applicants was low, no appointment should be made.[47]

Despite the fact that many departments did not have a single consultant, two-consultant departments were not long in coming. The first was in Leeds in 1974 when Michael Flowers joined David Wilson. The appointment process in this case has been described.[48] Once the moratorium was in place, the only way to appoint a second consultant was on the retirement of a MA as these had to be replaced by consultants. These appointments were not always a success as, unfortunately, personality and other problems started to come between colleagues. A&E was not like, for example, surgery where consultants had different wards, junior staff and nursing staff and at that time could (and frequently did) have totally different ways of managing patients. In A&E, junior staff, nursing staff, space and, usually, patients are shared, which means that teamwork and common policies are essential. Add to this, the very varied backgrounds of consultants and some of the inappropriate appointments described above and it was inevitable that problems would arise. Again, it is not possible to identify individual departments but in an article on multi-consultant departments, Wilson describes some of the problems of departments with second consultants[48] based on his experience of inspecting departments for the Specialist Advisory Committee (SAC). The senior registrar representative on the CSA Executive Committee (discussing the possibility of multi-consultant departments following the Short Report) wrote: 'Two consultants ... should be able to work out common patient management policies and should be able to sort out the problems of administration, clinical responsibility, teaching etc etc (though I gather that in some hospitals this has caused difficulties and maybe needs looking into before two consultants is accepted as a norm).'[49] David Wilson also wrote: 'If you are placed, as we are, working in close co-operation day in day out for a large part of our working lives, selecting your second consultant is a decision so crucial that I would place it only second, in order of importance, to selecting the person you marry.' Needless to say deciding who you are going to appoint before the interview is, now, not allowed! Problems of this type still occur but, probably, became less common once fully trained senior registrars started to be appointed.

Lewin felt that 'when there are two Accident and Emergency consultants in a department, it may be advantageous to have one medical and the other surgical'[15] and this was a commonly held view but with the vast majority of consultants having had a surgical training, this was rare. One place where it did happen was at the Royal Victoria Hospital, Belfast where Peter Nelson with a medical background joined William Rutherford. This idea became less common when doctors entered senior registrar training at an earlier stage without having done a number of years as a registrar in another specialty or came into the specialty with an Fellow of the Royal College of Surgeons of Edinburgh (FRCS(Ed)) in A&E. Senior registrars' training was designed to equip them with the skills in specialties which they had not previously experienced and they thought of themselves as neither physician nor surgeon but rather as an A&E trainee who happened to have a higher qualification in either surgery or medicine. Many is the 'surgeon' in A&E who has developed a major research or teaching interest in a medical aspect of the specialty (and vice versa).

Walpole Lewin's report was very supportive of A&E consultants but his was a one-man report done very rapidly to answer an urgent question. Walpole Lewin died in 1979 and the JCC wanted a more comprehensive report on A&E. A further working party was set up under the chairmanship of Mr R.H.B. Mills (an orthopaedic surgeon) 'to consider the recommendations in the Lewin Report and identify those areas in which progress towards implementation could be made and any other related matters as they see fit'. This reported in 1981 and is known as the Mills Report[38]. It too was very supportive of A&E (Howard Baderman was on the working party.) By this time the majority of A&E departments were being run by A&E consultants with 126 major departments run by A&E consultants and only 115 run by orthopaedic surgeons. Most A&E consultants by this time had a higher qualification and senior registrars were in post and emerging at the end of their training programmes. The Department of Health had indicated that they wanted to see an A&E consultant in every department but the Report still saw A&E consultants as just one way of running a department: 'It should be decided locally how best to provide the consultant services for an Accident and Emergency Department. Some will continue to make arrangements within their present departments along previously well-established lines in which the Orthopaedic Department continues to exercise control but we believe that even if this is so Consultants must have a well defined sessional commitment to the Accident and Emergency Department.' In smaller departments, it was thought that there was insufficient work and interest for a full time A&E consultant and it was felt that in such departments, consultants could have '… a firm but limited number of sessions in the appropriate specialty in which the candidate has previous experience'. But this should not 'undermine the essential principal that a consultant in an Accident and Emergency department should spend the majority of his time actually working in the department and we would consider that a minimum of 6 sessions is necessary to fulfil this fundamental duty'.

There was still great pressure to increase the numbers of A&E consultants and since the moratorium at least three new posts had been created by replacement of posts held by surgeons undertaking predominantly A&E work but categorised as General Surgeons. Mills felt that tight control on expansion should be exercised and that new posts should be justified by a qualitative and quantitative analysis of workload. He recommended that wide consultation with local consultants and College Regional Advisors was needed to ensure that the job description was adequate and that the fully documented case for the establishment of a new post should be presented to Central and Regional Manpower Committees. The Colleges should strengthen their advice to assessors on appointments advisory committees (AACs). If all that happened, he recommended that the moratorium could be lifted after a year.

4
Senior Registrars and Training

The JCC subcommittee which agreed in 1971 to appoint 32 A&E consultants as a pilot study recommended that 'no formal higher training programme in accident and emergency medicine be defined as yet but that preliminary consideration be given jointly by the Royal Colleges and other interested bodies to formulating suitable requirements, and that they be assisted in their survey by the supply of any information which becomes available from the operation of the pilot schemes'.[1] This reflects the Royal College of Surgeon's view that no higher training programme be set up unless the pilot scheme revealed a clear demand for the creation of a specialty of emergency medicine and surgery.[2] There were registrars in A&E departments but these were considered a service rather than a training grade and registrars were either on a surgical rotation or else were jointly appointed between orthopaedics and A&E.

However, one of the things dearest to the hearts of the early pioneers of A&E was how to train those who would follow them. This was, of course, essential for the future of the specialty but the interest was not entirely altruistic. Senior registrars and registrars are of enormous help to a department and to a consultant. More experienced than SHOs, they need less supervision and they help to supervise the SHOs. They are a shield that protects the consultant partially when an SHO wants to ask advice. Senior registrars intending to make a career in a specialty tend to be enthusiastic and contribute new ideas, teaching and research to a department. Leeds submitted an application and proposed a programme for senior registrar in July 1972[3] before many of the 30 original consultants had even taken up their posts. Derby did the same the following year.[4] As the first consultants were taking up their posts in the autumn of 1972, the CSA at their committee meeting in October started to discuss senior registrar training[5] and the following year they argued: 'A career structure in emergency medicine as a specialty should be available in the usual form of a graded ladder of registrars and housemen together with appropriate training both within and outside the department up to consultant level.'[6] They believed that trainees needed exceptionally broad experience extending to as many of the medical and surgical specialties as possible (with priority being given to orthopaedic surgery) and recommended experience at registrar level in both general medicine and general surgery with experience also in general practice, outside organisations (e.g. fire, police, ambulance, social services) and hospital administration. Unfortunately the President of the Royal College of Surgeons did not feel able to support any effort to draw up a definite training programme.[7] There was undoubtedly much opposition to setting up a training programme which David Wilson described in an article as being a rearguard action by those opposed to A&E as a specialty[3] though those opposing it would probably have said that they were awaiting the formal evaluation of the pilot posts as originally recommended. David Wilson also quotes a senior orthopaedic surgeon as saying, 'Consultants in accident and emergency should not be trained, they should be found.'

In October 1974 the Joint Committees for Higher Medical and Surgical Training (JCHMT and JCHST) set up a working party which reported in January 1975, unanimously recommending the need for a higher training programme and the setting up of a Specialist Advisory Committee (SAC) which would be answerable to both organisations.[8] This was initially not agreed by the Royal College of Surgeons but was eventually approved by the JCC in October 1975. The SAC had its first meeting on 3 December 1975 under the Chairmanship of JIP James, Professor of Orthopaedics in Edinburgh. There were representatives of the JCHMT, JCHST, Joint Committee for Higher Training in Anaesthetics (JCHTAnaes), British Paediatric Association (BPA), Royal College of General Practitioners (RCGP) and the CSA. The CSA representatives were David Wilson and Howard Baderman. The year 1976 was spent devising the curriculum and the criteria against which departments would be inspected and inspections of departments started in 1977.[9] These were not rubber-stamping exercises: a history of the Glasgow Royal Infirmary describes how, in 1979, the 'Special Advisory Committee' [sic] visited that hospital and 'was appalled at the conditions in the department, threatening to withdraw recognition of it as a suitable place for training postgraduate students'.[10] This was not the first and certainly was not the last hospital to be made to upgrade its A&E department by the SAC or its successor, the Joint Committee for Higher Training in Accident and Emergency (JCHTA&E).

The entry requirement to higher training was wide general experience and a higher qualification (MRCP, FRCS, FFARCS or MRCGP were equally acceptable). The training was to be four years, with one year working in a busy A&E department, two years of working attachments in related specialties in which the SR did not have previous experience and a final year taking increasing responsibility in a large A&E department.[11,12] Retrospective recognition was given for previous training which was considered of relevance and done after obtaining a higher qualification.

The new posts had to be funded, approved and inspected and Lewin reports that the first senior registrar took up post in January 1978.[1] The Report is, however, contradictory as appendix E, a letter from the Chief Medical Officer indicates that there were senior registrars as early as 1973. His estimate of SR numbers is shown in Table 4.1.

One of these senior registrar posts in 1973 was probably William Adams (see below) but the others must, I believe, have been the result of confusion between orthopaedics and A&E. However there were senior registrar posts in A&E before 1978. The CSA Committee minutes of October 1973 record that University College Hospital (UCH) had advertised for Senior Registrars but that the wording of advertisement was such that no satisfactory replies were received and no appointment was made.[4] Later (probably at the end of 1973) William Adams was appointed as a 'pilot' senior registrar at UCH. He left in 1975 or 1976 but the post was never evaluated and he was not replaced until the SAC started approving posts.[13] In addition there were senior registrars in A&E in Scotland. Alasdair Matheson was appointed senior registrar in Aberdeen in 1975[14] and Stuart Lord was appointed at the Glasgow Royal Infirmary in 1976.[15] When he obtained a consultancy in 1977, he was replaced by Ian Swann. The consultant at that time was Mr Simpson

Table 4.1 A&E senior registrar numbers

Year	Number
1973	3
1974	4
1975	4
1976	4

Source: Lewin Report, Appendix E.[1]

who was not full time in A&E as he combined this with hand surgery (and later became a full time hand surgeon). At the same time William Morgan was an SR at the Western Infirmary, Glasgow. These posts were not recognised by the SAC but presumably were recognised by one of the Scottish Royal Colleges. David Ferguson was appointed in Belfast in August 1977.[16] I am uncertain as to how this happened and William Rutherford (his consultant) cannot remember but I suspect that Mr Rutherford got fed up with waiting for bureaucratic wheels to turn and approve posts and 'jumped the gun'. William Rutherford himself says that such behaviour would have been most unlike him![17]

In Manchester, Prof. Miles Irving, the professor of surgery had a major interest in trauma both from the academic and teaching perspective. He was keen to develop the academic potential of Hope Hospital, Salford which had become part of the expanding Manchester University Medical School and which was one of only two hospitals in the North West Region to have an A&E consultant. He brought together Gordon Laing from Hope Hospital and Malcolm Hall (Preston) to form one of the first senior registrar training programmes in England. David Yates was appointed tutor (and honorary senior registrar) in A&E in 1976 within the department of surgery.

It was obvious that the first SAC-approved senior registrar posts were going to be advertised and one or two individuals were preparing themselves. David Wilson had set up a rotating registrar post in Leeds which was filled by Andrew Marsden who obtained the first SR post in Leeds in January 1978 with the first NHS posts in Manchester being filled very soon afterwards.

By the time the Lewin Report was written in 1978, there were 18 posts with manpower approval and ten of these had been approved by the SAC with four SRs in post in England and Wales and three in Scotland.[1] The first senior registrars, being the sole SRs in their hospital and frequently their region, lacked support. In 1979 six of them met in the Lake District and formed the A&E Senior Registrars Travelling Club (which became the British A&E Trainees Association in 1990) and this appointed representatives to various CSA committees. SR numbers did not take off until about 1980 but by February 1981 the Mills Report states that there were 26 SRs in post. Ten were in their first year, 15 in their second and one in his third year. A further 11 had been appointed to consultant posts and two had left the specialty.[18] Apart from these 26 posts, there were a further 12 awaiting advertisement and five which had been approved but which were unfunded. Thus 43 posts had been approved. Of the senior registrars in post, about 80 per cent were British

Table 4.2 Higher qualifications of the senior registrars in post February 1981

Qualification	Number
FRCS	19
MRCP	2
MD	1
FFARCS	2
MRCGP	2

Source: Mills Report.[18]

graduates and most had a surgical higher qualification. The qualifications of the 26 in post are shown in Table 4.2.

The high percentage of A&E SRs with an FRCS reflects those who applied. In the Yorkshire Region from 1977 to 1983, 89 per cent of applicants for SR posts had a FRCS as their sole higher qualification.[19] Just as in the 1970s doctors who had failed to make progress in other specialties tried to obtain consultant posts in A&E, so in the early 1980s they tried to obtain senior registrar posts in the specialty. The Department of Health Statistics show that in 1980/81 there were 12.8 candidates for every A&E SR post compared to 3.7 for all specialties combined. The mean age of appointment to an A&E SR was the oldest for all specialties at 33.5 years (all specialties combined 31.9 years).[20] In 1981/82 there were 10.8 candidates per post (all specialties 4.1).[21] Most would have been appointed from registrar or senior registrar posts in other specialties.

The problems of the early SRs were described in an article by Stephen Miles, a senior registrar in London.[22] As with the early consultants, 'most of the SR recruits have tended to be people who have progressed to a fairly high level in a specialty before deciding on this career' as there was no A&E training for them to have aimed for. The first SR in a department had to develop a role for himself and did this by introducing additional teaching, follow up clinics, 'quality control', research and being available to give second opinions. This enhanced the activities of the department but when the SR left to do long secondments, the department gradually reverted to its former level so when the SR returned, he had to start again from scratch. The early SRs were usually very experienced in one of the major specialties and found it difficulty doing attachments at SHO level. Sometimes the team to which the senior registrar was attached also found it difficult to cope with an 'SHO' they had previously sought advice from as an A&E SR. The SRs argued that for the person wanting to make a career in A&E from the outset, this experience of other specialties should be done at registrar level. They also believed that the programme as it existed did not properly address management skills.

The possibility of a specialist diploma in A&E had been raised at the CSA at least as early as 1974[23] but the Lewin Report in 1978 advised that the '...dangers of fragmentation have to be borne in mind. In any event, it would seem premature to discuss diplomas before the specialty is firmly established.'[1] However that same year discussions began at the Royal College of Surgeons of Edinburgh between Myles Gibson a neurosurgeon and member of College Council and Keith Little an A&E consultant in Edinburgh. By October 1981 final details of the Edinburgh FRCS in Accident and Emergency Medicine and Surgery were announced to the CSA committee. 'The committee received this news with acclamation and William Rutherford, the president, described it as a significant milestone in the development of the specialty of A&E medicine.'[24]

The senior registrars were less enthusiastic. Their main concern was probably a fear that they might be expected to take the exam themselves but they also had other concerns. Around the same time, the Royal College of Surgeons of Edinburgh had introduced specialist exams in orthopaedics and neurosurgery to be taken at the end of training. The A&E exam, taken to enter training, did not compare. There was also concern that it would not be recognised as equivalent to the MRCP or FRCS especially if honorary diplomas were given to existing specialists as had happened with some other higher qualifications. (This did not happen.) The number of people wanting to do the exam would be small and it would take a lot of time to organise and the senior registrars thought that the time and energy of candidates and examiners could be used better by improving the service given in A&E, improving teaching to junior doctors and students and doing research. They felt that it might be reasonable to set up a diploma in about 20 years when the specialty should be better established.[25] The exam was also not well received by the Royal College of Surgeons of England. The first sitting of the exam was in May 1982 with six candidates. Three passed of whom two were SRs. The exam was accepted by the SAC as an entry qualification to higher training.[26]

In 1978, Lewin investigated the 77 A&E registrars and found that only 11 were on the establishment of an A&E department under the supervision of an A&E consultant, 13 were on a surgical rotation and 38 were orthopaedic trainees with a variable A&E involvement. The rest were obscure. He recommended that A&E registrars should not be part of the

A&E establishment but should come to the department on rotations.[1] This meant that A&E lacked the career structure of other specialties and doctors intending a career in it had to obtain a registrar post in some other specialty first. This was said to be one of the reasons for the specialty's unpopularity.[27] This was rectified by the Mills Report which recommended that a limited number of registrar posts should be established to bring the career structure more in line with other specialties but that most registrar posts should still be in rotations with the other major disciplines.[18]

In 1983 it was recommended that the MRCGP be dropped as a recommended entry qualification to training in A&E. Most within the specialty did not consider MRCGP inappropriate but doctors trying to enter senior registrar training with it as a sole higher qualification found that some people, particularly in other specialties and including some Postgraduate Deans, did not see it as the equal of the other higher diplomas and so candidates with this as their sole higher qualification were disadvantaged.[28] This decision upset some of the A&E consultants who had entered the specialty with an MRCGP.

Further information on the adequacy of early SR training is obtained from an article based on a questionnaire sent to the 48 consultants who had been appointed between September 1982 and July 1987 and who had been on a training programme. There was a 96 per cent response rate. Ten had spent at least 3.5 years as SR.[29] Most thought that their training in general clinical problems and in trauma, cardiac and paediatric resuscitation was adequate but there were criticisms of their training in eye, ENT, obstetric and gynaecological problems and management training was said to be poor. (Interestingly eleven years later another survey was done of senior registrars that found that most were happy with their training but that the deficient areas were almost exactly the same.)[30]

An interesting part of the 1988 paper also describes the difficulties which this generation of consultants had after their appointment. These are shown in Table 4.3. Those in post as senior registrars for over 3.5 years had fewer problems than those in post a shorter time.

As late as 1991, the majority of A&E registrars still had a surgical higher qualification, often in addition to the Edinburgh A&E exam as many felt that the FRCS(A&E) as a sole qualification was not sufficiently recognised in A&E circles or by other specialties and lacked credibility against MRCP and FRCS.[31]

It would be of great interest to determine what an A&E trainee actually does. I am not aware of any description of the work of a senior

Table 4.3 Problems encountered by SRs after becoming consultant (affecting over 50 per cent)

Problems	Percentage
SHO numbers	72
Equipment shortages	69
Financial constraints	69
Secretarial problems	65
Nurses (numbers and attitudes)	65
Management/administration	56

Source: Reference 29.

Table 4.4 The work of an A&E registrar in a year

New patients seen as principal doctor	3,251
Verbal opinions given for an A&E doctor	511
Radiological opinion given for an A&E doctor	805
ECG opinion given for an A&E doctor	76
Opinion given after seeing the patient	474
Review of patient recalled after receipt of XR report	16
Opinions given for other specialist teams	218
Patients reviewed on short-stay ward	256
Patients reviewed in general-return clinic	294
Patients reviewed in hand clinic	131
Patients reviewed in dressings/wounds clinic	115
Telephone advice to GP	32
Telephone advice to patients	84
Cardiac arrest	46
Other medical emergency	149
Significant trauma	72
Other significant surgical emergency	16
CVP line insertion	42
Insertion of ETT	14
Manipulation of Colles fracture	70
Biers blocks	46
Insertion of chest drain	13
Reduction of shoulder dislocation	36
Reduction of other dislocation	5
Diagnostic peritoneal lavage	3
Repair extensor tendon	3
Insertion of Sengstaken tube	1
Informing family of the death of a relative	16
8.7% of new patients were critically ill	

Source: Reference 32.

registrar but Jonathan Wyatt wrote an interesting paper describing his work as an A&E registrar in Glasgow in 1992/93.[32] This is summarised in Table 4.4. This is the work of one person (clearly obsessional to note every patient contact for a year) in one hospital at one time and one cannot be certain that it is typical but it demonstrates the role of a registrar in supervision and giving advice.

By the mid-1980s more registrar posts were under the direct control of A&E consultants and it became expected that applicants for SR posts would come from the ranks of A&E registrars rather than transferring from registrar posts in other specialties though many still did a registrar post in another specialty before moving to A&E. Registrar posts provided valuable experience but were still seen as a way of providing a service rather than training. If a department had too few registrars to provide middle grade cover, another way of achieving this was to appoint experienced SHOs to work as registrars and on the same rota. These were known as SHO3s and were particularly common in Scotland. This role had no official standing and the posts were only recognised for training at SHO level though middle grade experience would have been looked on favourably by appointments committees.

In 1991 training in A&E changed so that it commenced at registrar level and trainees did three years as a registrar followed by two as an SR.[33] Following the Calman Report these two grades were amalgamated into the specialist registrar (SpR) grade as from 1 July 1996.[34] This caused little difficulty in A&E as the posts combined easily to a five year SpR post. The Calman Report also formalised the training. A syllabus was produced and trainees were to have regular appraisal with an annual review. Their progress would be monitored with regular assessments and the end of training would be marked by obtaining a Certificate of Completion of Specialist Training (CCST) which would allow them onto the specialist register. To obtain a CCST requires passing the Fellow of the Faculty of Accident and Emergency Medicine (FFAEM) exam and passing an end-of-programme assessment. Since 1997, only doctors on the specialist register have been able to obtain a consultant post.

With increasing popularity of A&E as a career, entry into training posts has become more competitive. While the training programme allows secondments in 'essential specialties' (acute medicine and cardiology, paediatrics, orthopaedics, anaesthesia and intensive care and surgical specialties) it is now uncommon for a person to obtain a specialist registrar post unless they already have SHO experience in most of these specialties and unless they already have some middle-grade experience either as a locum specialist registrar or as a staff grade (SG) doctor or clinical fellow.

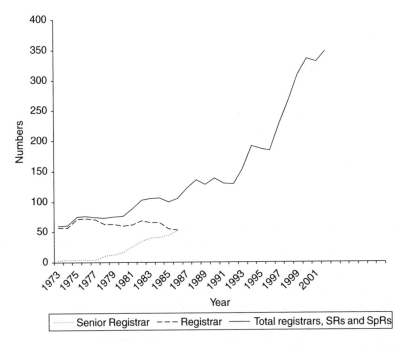

Figure 4.1 Number of registrars, senior registrars and specialist registrars in A&E in England 1973–2002
Source: www.doh.gov.uk/stats/history.htm

One of the major themes in the history of A&E since 1972 is the number of consultants required in the specialty. From this can be worked out the number of registrars required to fill future consultant posts. This is explored in Chapter 5. The actual numbers of senior registrars, registrars and specialist registrars is shown in Figure 4.1.

With changes in surgical training, the FRCS (Ed) exam in A&E became MRCS (Ed) and this is now the most popular entrance exam into the specialty. (The FRCS (Ed) continued until 2003 for foreign doctors.) Once the Faculty was formed, it seemed inappropriate that entry to the specialty was by exams set by other Royal Colleges and in 2003 the Faculty started its own exam to allow entry to Specialist Registrar training in A&E – the Member of the Faculty of Accident and Emergency Medicine (MFAEM) exam.

5
How Many Consultants?

Deciding how many consultants are required in a specialty is obviously essential for the provision of the service but is also vital in order to plan the number of training posts needed to fill future consultant vacancies. If too few are trained, consultant numbers cannot expand or posts may be filled with untrained individuals but if there are too many trainees, there is a risk of unemployment. Because the length of training is four to five years and financial planning is done a year in advance, it is necessary to try to predict requirements six or seven years into the future. This has been a major cause of difficulty over the years (in all specialties.) Although at times there have been fears that there may be too many training posts, the problem in A&E has always been too few trained individuals to fill consultant posts. In part this may be because recommended consultant numbers seem to increase when there are plans to reduce the number of trainees: at times the tail (trainee numbers) appears to wag the dog of consultant numbers.

For approximate calculations, it was usually assumed in reports that consultants would be appointed at the age of 35 and would remain in post for 30 years. Senior registrars (SRs) were in post for four years and so a ratio of one SR to 7.5 consultants was required or 1 : 8 to accommodate some wastage. Registrars usually spent two years in post and so the ratio of one registrar to two SRs was needed (with a small allowance for wastage). Specialist Registrar training, introduced in 1996, was five years in duration and so a ratio of 1 : 7 was required. This is not an essay on manpower planning but adjustments obviously need to be made for the age distribution of consultants and changes such as a tendency towards early retirement or part-time working.

Within a year of the 30 consultants being appointed in the pilot study in 1972, the CSA had specified its ideal staffing levels in a document on providing an integrated emergency service: 'For a department dealing with at least 25,000 new patients per annum and providing continuous 24-hour cover, the minimum staff should be one consultant and four or five senior house officers. The ideal staffing for such a department would be two consultants, one senior registrar and five to six senior house officers.'[1] By 1980 it had produced its first document on staffing norms which recommended a consultant for all departments seeing more than 20,000 new patients and two consultants for departments seeing more than 30,000. For departments seeing more than 30,000 new patients per year it recommended to 'extrapolate to scale'[2] but I believe that this extrapolation referred more to other grades of staff as very few people at that time would have recommended five consultants for a department seeing 75,000 patients per year.

The Lewin Report, in 1978, made no recommendations on consultant numbers but just noted that the ideal number of A&E departments was about 250 and so commented that if most departments had an A&E consultant, 'it would require about 250 consultants in England and Wales'.[3]

The following year an article in the *BMJ* said that there was still controversy about who should run A&E departments[4] with different groups recommending orthopaedic surgeons, accident surgeons and casualty consultants and others recommending flexibility and there were still different views on the role of the casualty consultant. Even by 1981, there was no agreement that all departments should have an A&E consultant and the Mills Report was 'not certain of the pattern of staffing which will evolve on the retirement of some Trauma and Orthopaedic Surgeons with a specific sessional allocation to Accident and Emergency Departments'. However the DHSS had said informally that they would anticipate one A&E consultant in every DGH and this would indicate the need for approximately 250 consultants. Some very large A&E departments would require more than one consultant and it was assumed that the final number would approach 300. This would require 30 SRs and Mills advised that great care must be exercised in further expanding the number of SRs.[5] The Chief Medical Officer's (CMO) response was that Mills gave no indication about how a department should be covered out of hours and that it was unreasonable to expect a consultant to work more than one in three rota. Alternatives would be for departments to have more than one consultant or for other consultants to cover.[6]

Thoughts within the specialty were considerably influenced by external pressures. There were major problems with medical manpower planning in the UK with far more junior doctors than could be accommodated in career posts. The DHSS launched a manpower initiative in June 1980 which was sent to chairmen of Regional Health Authorities in September 1981. It recommended doubling the number of consultants in 15 years with a ratio of 1.8 consultants to one junior doctor. This was flawed in that it assumed no growth in GPs and made no allowance for doctors making careers outside hospital and general practice. Their projected figures had also reduced SHO posts below the number required for 4000 new graduates and hence would have guaranteed medical unemployment.[7] Despite this, Devon Area Health Authority (and doubtless other authorities) projected the future staffing of their A&E departments by doubling existing consultant numbers and estimating junior doctor numbers (house officers, SHOs, registrars and SRs combined) using the 1 : 1.8 junior : consultant ratio. Their projection is shown in Table 5.1 and this would obviously be inadequate to either provide a service or 24-hour cover.

Table 5.1 Projected staffing of A&E departments in Devon in 1981

	Current (WTE)		Planned (WTE)	
	Cons	Juniors	Cons	Juniors
Exeter	1	6	2	1.11
Torbay	1	4	2	1.11
Plymouth	2	8	4	2.22

Note: North Devon District Hospital, Barnstaple which did not have an A&E consultant at that time is not mentioned.

Source: Devon AHA. Report on Hospital Medical Staffing Structure 13 November 1981.

The following year the Short Report on medical education[8] was published. It accepted the ratio of one junior to 1.8 consultants (across all specialties) and argued this for quality of care rather than just manpower planning. It felt that 'patients are entitled to be treated by trained specialists (i.e. consultants) ...' and that

> a much higher proportion of medical care should be provided by fully trained medical staff than at present. ... Where junior doctors are treating a patient, they should be under the close supervision of a specialist.
>
> It follows that in most hospitals and in most specialties there should be an increase in the number of consultants and a decrease in

the number of junior doctors. ... The nature of consultant work will change and the consultant will no longer be able to rely on junior doctors to perform basic routine or emergency work. However, with more consultant colleagues, each consultant will have fewer patients and this should allow a much closer doctor-patient relationship with greater continuity of care. ... While consultants as a whole will have to undertake a greater share of the emergency, out-of-hours, work there will be an increased proportion of young consultants and with good will and flexibility it should be possible to avoid the burden of work falling on older consultants. ...

Consultant's time certainly costs more in salary terms than junior time but consultants function more efficiently and may therefore provide less expensive care. A consultant provided service could result in higher productivity and better value for money as well as improving the quality of care. This higher productivity might well reduce waiting lists and the costs of continuing sickness, although incurring higher hospital expenditure.[8]

The Report also argued for more general professional training: 'We, therefore, recommend that all specialties, through their Colleges and Faculties, should require post-registration trainees to spend periods totalling at least one year in other disciplines.' This should be to get broad experience. 'It would not be sufficient, for example, for a general surgeon to spend a year in orthopaedics.'

The Short Report did not specifically mention A&E and in the manpower statistics quoted in the report, A&E was lumped together with orthopaedics (a common, if irritating, occurrence in the early 1980s.) The emphasis on general professional training would clearly benefit SHO staffing in A&E departments but all the talk was on consultant numbers. In 1982 when only a handful of A&E departments had two consultants and many had no consultant, Vera Dallos, A&E consultant at Whipps Cross Hospital and Chairman of the A&E Subcommittee of the CCHMS asked the CSA Executive Committee for their views on A&E consultants working a shift system.[9] A CSA working party was set up and recommended that all departments should have two consultants and that staffing needs for small and large departments differed. It agreed that SHO numbers should be maintained and that any increased consultant numbers should come from a reduction in middle grade numbers. Three members of the working party including the President and President Elect felt that it was desirable to provide consultant cover at all times in the department and that shift work was unavoidable.

Two members including the senior registrar representative disagreed strongly.[10] (With such a difference of opinion, the conclusions of the working party were meaningless.) One of the problems with medical politics is that decisions with long-term implications are made by senior members of the profession who will not be in post when the results of their decisions come to fruition and the senior registrars were not impressed. The senior registrar representative wrote to Dr Dallos:

> We were all horrified by the fact that three very senior and influential members of the CSA were advocating continuous consultant presence in the department. We would like to ask whether these three would personally be prepared to forgo all the status and responsibility which they have fought for and so rightly achieved in order to work under the direction of another doctor. ... Senior registrar training is directed towards, not only our clinical skills, but also towards organising an A&E service for a hospital and district. ... This forms a substantial part of a consultant's workload and is an aspect of the work to which the senior registrars are looking forward and which most consultants would be reluctant to give up. ... The senior registrars agree with the Short recommendations of doubling the number of A&E consultants to a norm of two per department but the profession has not been asked to increase consultant numbers beyond this and we would oppose any attempt to.[11]

In fairness, not all senior registrars felt like this as, at an earlier meeting, a small number (who presumably did not attend the later meeting) had indicated that they would be prepared to work shifts.[12]

The senior registrars were not alone in opposing the idea of working shifts. There was an acceptance that senior cover was needed out of hours in the A&E department but dispute as to how this should be achieved. At the CSA AGM in 1980, some members favoured an expansion in the consultant grade while others wished to see the introduction of a non-registrar intermediate grade.[13] Regional CSA meetings, called to discuss the Short Report, recommended middle-grade cover until midnight and in some cases 24 hours but not necessarily consultant presence in the department for all this time. In general terms, large departments wanted multi-consultant departments with extended cover while smaller departments wanted a single consultant.[14]

The crux of any discussion on consultant numbers depends on the function of the consultant. The Lewin Report[3] had defined that the functions of an A&E consultant are:

- To organise and administer the department,
- to provide an initial diagnostic service at high level, and arrange for further care if necessary,
- to provide emergency resuscitation,
- to provide teaching for those working in the department,
- and to maintain liaison with the ambulance, police and fire services, and with the community, and to take part as required in teaching programmes for their staffs, and first aid training for the public in association with the recognised voluntary organisations.

This job description is for a job combining administration and teaching with resuscitation and whereas SRs and consultants realised that they had to deal with less serious cases, and they had not signed up for a job involving shift work. Almost all consultants and senior registrars would have spent several years as a junior doctor on a one in two (or even more onerous) rota with the need to be resident on call and during this time they would have looked on a consultant post as a light at the end of the tunnel. The Short report threatened to extinguish this light.

The senior registrars need not have worried as in March 1983 there were 15 vacant consultant posts[15] and there was no prospect of enough consultants to make shift work happen.

In 1983 the DHSS expected consultant expansion over the following few years to a target figure of 300. This would require 40 SRs and 20 registrars. On this basis most departments would not have middle-grade training doctors.

At the CSA AGM in 1984 William Rutherford, the President, said that it would be interesting to evaluate the experience of America where it had been found cheaper to employ experienced emergency physicians because of the rise in standards of care and reduced likelihood of litigation.[16] In a journal article he later expanded his views in the context of improving care for victims of trauma: 'I would be happy to see accident and emergency departments staffed to a level where at least in one hospital in every city round-the-clock senior cover on site was available. I see no immediate prospect of staffing on such a level. The next option would be that all the senior accident and emergency staff for a city, regardless of their hospital affiliation, should form a single service.' All would do out-of-hours duties in one hospital or there would be a different designated hospital on call every evening. If this could not work, his next option was the formation of a roster of consultants drawn from A&E and orthopaedics.[17] This occurred in Belfast during the Northern Ireland troubles when consultants in different specialties formed a rota for sleeping in the hospital to manage major trauma. Mr Rutherford as an A&E consultant at the Royal Victoria Hospital in that city had taken part in this rota.[18]

As noted, the CSA in 1980 had recommended two consultants for a department seeing 30,000 new patients per year but this was widely seen as being unachievable in the foreseeable future and the practical threshold for two consultants was variously set in different CSA documents at 40,000,[19] 45,000[20] and 50,000.[21] By 1984 there were about 150 consultants and 269 departments in the UK seeing more than 20,000 new patients, 112 of which were seeing more than 45,000 new patients. There was therefore a need for 381 consultants and 36 SRs. Consultant expansion (across all specialties) had failed to meet the 4 per cent target and there were already 56 A&E SRs in the NHS with a further two in the Royal Navy. It was recommended that SR expansion be stopped and that numbers would need to be reduced.[20]

There was still debate as to what to do with departments seeing fewer than 20,000 new patients per year but the 1984 CSA AGM voted that all DGHs should have an A&E consultant.[20] This was, of course, an A&E decision but there was still not a unanimous view within the medical profession that all departments should have an A&E consultant. In 1985 a list was produced of hospitals without an A&E consultant.[22] This included 15 hospitals seeing more than 40,000 new patients per year. The largest was the North Staffordshire Infirmary in Stoke (77,900 new patients per year) and the list also included two teaching hospitals (Bristol Royal Infirmary and the John Radcliffe Hospital, Oxford). Some hospitals may have failed to attract an A&E consultant, in some there was no opposition to A&E but it was just not high on their list of priorities but other hospitals still opposed the concept of A&E consultants.

The ratio of one SR to eight consultants supposes a stable number of consultants but A&E consultant numbers continued to increase and the problems of the late 1970s persisted in that senior registrar numbers were inadequate for the consultant expansion required. Senior registrars could therefore be choosy about the jobs they applied for. Many were still applying for posts long before the completion of their training[23] but others could wait and have their pick of jobs. There was still concern about consultant job descriptions[24] and one, for a post in Manchester was described as 'outrageous'.[25] These jobs found it hard to get suitable applicants. Other jobs rejected by SRs were those where it was feared that orthopaedic surgeons would want the A&E consultant to be subservient. Soaring house prices in the South East made non-teaching hospital jobs in those areas less attractive.[26]

The Lewin Report had said that there was general support for consultants in A&E as long as 'in the future consultants ... [are] properly trained for the job'[3] but with inadequate SR numbers, inappropriate consultant appointments continued. In 1984 an orthopaedic surgeon noted: 'There is still wide dissatisfaction about the type of medical practitioner being appointed to A&E departments'[27] and this view was accepted by the CSA.[28] In 1989 the President of the Royal College of Surgeons expressed concern over A&E consultant appointments when there was no SR applicant.[29] The logical solution was only to appoint from the senior registrar ranks. The editor of the *British Journal of A&E Medicine* agreed: 'Another situation in which our specialty devalues itself is the practice of appointing doctors untrained in accident and emergency medicine to positions of seniority. ... While it was once acceptable for retiring members of the armed forces or the missionary service, with their wealth of general experience, to enter accident and emergency medicine as consultants, that time is now passed. If suitable senior registrars are not applying for the posts, it would be better that the posts were allowed to stay vacant.'[30] Hospitals (and single-handed consultants) did not always agree, believing that, sometimes, anybody was better than nobody.

A number of things combined to fuel the demand for more consultants. Hospitals without A&E consultants were increasingly learning of their value from other hospitals and surgeons who had opposed A&E, were retiring and being replaced by younger surgeons who had seen the benefit of A&E consultants during their training and did not want the responsibility of running A&E departments themselves. Single-handed A&E consultants wanted a colleague to share the responsibility for the department as increasingly it was felt that other specialties covering A&E was inappropriate. An editorial in the *British Journal of Accident and Emergency Medicine* stated: 'If, as I fear, our specialty is still held in low regard by some members of our profession, we have to take our share of the blame. Too often we accept cover for our department from other disciplines. ... It is no more appropriate for an orthopaedic surgeon, for instance, to cover an accident and emergency department than for an accident and emergency consultant to be on call for orthopaedics.'[30] The feeling was mutual as, increasingly, other specialties did not want or feel able to cover A&E. While other specialties would cover (willingly or otherwise) a single-handed A&E consultant, once a second consultant was appointed, this offer of help frequently ceased. With six weeks annual leave and ten days study leave per year, a consultant working with a single colleague will be single handed for more than seven weeks a year and this increasingly became unacceptable and so two-consultant departments looked to appoint a third. If a small survey of eight A&E consultants is to be believed, A&E consultants were working hardest of all specialties working an average 54.9 hours per week compared to the average for all whole and maximum part-time consultants of 49.2 hours.[31] This was accepted by the National Audit Office: 'The average working week of an A&E consultant exceeds 50 hours, the highest of any hospital specialty.'[32]

Another attempt to correct the imbalance between training and consultant posts was 'Achieving a Balance' published in 1986. This did not specifically mention A&E but its aims of consultant expansion and of newly appointed consultants in acute specialties accepting greater involvement in acute patient care[33] influenced the whole profession. The Royal College of Surgeons' report on patients with major injuries[34] in 1988 also argued for increased A&E consultant numbers.

The non-clinical workload of all consultants was increasing with medical audit, audit committees, appointments committees and other committees related to junior staff, other district committees and so on[35] and as A&E consultants started to become more senior within their hospitals, many began to take on additional responsibilities for education (e.g. clinical tutor) or within management. The late 1980s also saw the expansion of Advanced Cardiac Life Support (ACLS) and the introduction of Advanced Trauma Life Support (ATLS) courses in the UK and later, Advanced Paediatric Life Support (APLS) courses were introduced. These three-day courses are recognised as being excellent but are demanding on instructor time. Many A&E consultants became very involved with these courses and instructing on two ATLS and two ACLS courses per year would take 12 working days. All of these additional responsibilities necessitated additional consultants.

In 1991 there was action to reduce junior doctors' hours[36] and in 1992, 25 new, centrally funded, A&E consultant posts were established in England as part of the junior doctors hours initiative.[37]

Despite attempts at expanding consultant numbers, by 1991, 44 A&E departments seeing more than 20,000 new patients still did not have an A&E consultant and of the remainder most had a single consultant. Twenty posts were unfilled because of lack of suitable applicants.[38] These numbers do not include those not even advertised because a hospital knew that it would be unable to fill them. Around the same time, the Royal College of Surgeons decreed that it would recognise A&E departments for training only if they were led by an A&E consultant. This was a mixed blessing. In one respect it was recognition by the College of the

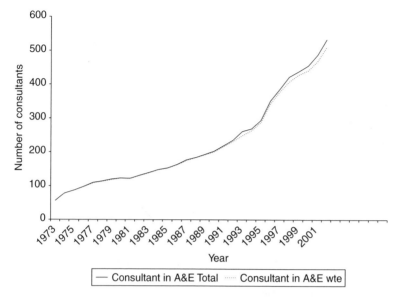

Figure 5.1 Number of consultants in A&E in England 1973–2002
Source: www.doh.gov.uk/stats/history.htm

specialty and its value but it led to a scramble by hospitals without a consultant to appoint one quickly for fear that its A&E SHO posts would lose training recognition by the Royal College. Thus more appointments of untrained people were made.

The increase in consultant numbers from 1973 to 2002 is shown in Figure 5.1. Consultant numbers in other specialties grew as well but A&E grew faster than other specialities as shown by Figure 5.2 which shows A&E consultants as a proportion of total consultant manpower over the same period.

In 1992 SR numbers were increased from 65 to 105 and registrars from 64 to 100.[39] Despite the desperate need, the posts were slow to be filled. New posts were frequently allocated in March for the financial year starting the following month. There was no central funding and hospitals had allocated their budgets for the year many months earlier and so these posts were difficult to fund. Hospitals were sometimes reluctant to fund secondments to other specialties (essential in A&E training programmes) as they felt that they were not getting anything for their money. Finally with the expansion in SR numbers and registrar posts being accepted as higher training, the workload of the SAC probably tripled and it had difficulties processing all the paperwork and doing the

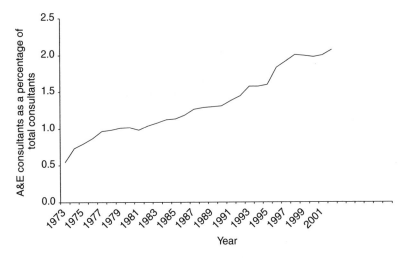

Figure 5.2 A&E consultants in England as percentage of total consultants 1973–2002
Source: www.doh.gov.uk/stats/history.htm

necessary inspections. By April 1993 virtually none of the additional allocated posts had been filled and in the preceding year 68 consultant posts were advertised but only 25 SR posts.[40] Not all of these 68 posts would have been filled but it is of interest to compare this figure of posts advertised with only 50 SRs accrediting between 1991 and 1994.[41]

The BAEM revised its recommended staffing levels in 1993[42] as shown in Table 5.2.

Table 5.2 BAEM consultant staffing recommendations 1993

New patients	Number of consultants
25–50,000	2
50–75,000	3
75–100,000	4

Source: Reference 42.

Two years later there were 267 consultants but using the recommended figures there was a need for 612 and so there was still a shortfall of 336. With 105 senior registrars, this shortfall would take 15 years to be corrected.[43] SR numbers were increased and in 1996 the SR and registrar grades were merged into a single grade of Specialist Registrar (SpR). Recommended numbers of SpRs continued to be increased year by year by the Specialist Workforce Advisory Group (SWAG) but no funding came with these recommendations and in 1996 it was reported that 7 per cent of A&E SpRs were unfunded. There were still problems funding the essential secondments outside A&E with at least one trust threatening to withdraw funding to SpRs for full time secondments.[44]

Following the Calman Report and legislation, it became a legal requirement that after 1 January 1997 doctors had to be on the specialist register before taking up a consultant post.[45] This led to a scramble during 1996 to beat the deadline by hospitals that would have found it difficult to attract a trained doctor, to appoint an A&E consultant. In 1996 there were 71 consultant appointments made. Fifteen were consultants moving from one hospital to another and 37 were of senior registrars (not necessarily accredited). In addition three Associate specialists (AS), five staff grade doctors and one registrar were appointed consultants, as were ten doctors who gave their previous grade as locum consultants. (It is uncertain how many of these were accredited senior registrars and how many had no specific A&E training.) In addition 27 AACs had to be cancelled because of insufficient applicants and 11 AACs made no appointment. The 71 appointments made in 1996 needs to be compared to 26 appointments made in 1997, 42 in 1998 and 35 in 1999.[46]

It was hoped that the requirement to be on the specialist register would finally put a stop to inappropriate appointments but in 1998 concern was raised when a person with a CCST in a surgical specialty was appointed to a post as a consultant in A&E – the legal requirement to be appointed a consultant is to have a CCST or to be on the specialist register but there is no law to say that you can only get a consultant post in the specialty in which you have trained![47]

Increasing SpR numbers to cope with the shortfall of A&E consultants continued until 1998. At that time it appeared that obstetrics and gynaecology was overproducing SpRs and their expected increased consultant expansion was not happening. There was therefore the prospect of unemployment within that specialty. This led to government concern and an insistence that that should not be allowed to happen in other specialties and in 1999 the recommended increase in A&E SpRs was cut from 30 to 10 by the Specialist Workforce Advisory Group (SWAG).[48] Later the same year the NHS Executive noted that there were 411 A&E consultants and 329 SpRs. With an expected 33 retirements over the following years, there would be 726 consultants by 2005. Even with increasing workload, the BAEM recommendations pointed to a need for about 800 consultants. Assuming 7–14 retirements a year, there was only a need for 35–70 SpRs and so there needed to be a reduction in the early years of the next decade of 260 to 294 SpR posts.[49] This could not be done in one blow and they recommended a gradual reduction. An immediate reduction of 75 posts was mentioned in the medical press.

There is no doubt that this would have caused havoc in staffing A&E departments which had come to rely on middle-grade staff to provide cover until midnight or even for the full 24 hours. A number of departments already had more than the BAEM-recommended number of consultants and many departments were planning further appointments to allow more out-of-hours consultant cover (see Chapter 6) but these arguments were countered by the NHS Executive who argued that the obstetricians and gynaecologists had planned consultant expansion but that this had not happened because Trusts had not funded the posts. However the decision to reduce posts was postponed for 12 months.

In 2000 and 2001 there was further discussion within the specialty on consultant numbers based on expanding the role of the consultant. If, in larger departments, one aimed for consultant presence in the department for twelve hours per day, seven days a week while continuing existing work (teaching, follow up clinics, ward rounds, administration etc) a department would need 8–10 consultants. A smaller department would need a minimum of three consultants. Routine work out of hours would require those doing such work to be paid a premium. This would require 1350 consultants by 2010.[50]

While these numbers have not been formally agreed, there is, again, a realisation that there are too few A&E consultants. Quite apart from too few consultants to fulfil their traditional role, there is increasing pressure for consultants to take on additional work such as medical emergencies and care of minor problems (see Chapters 6 and 11). Some consultants want to work part-time either because of domestic commitments or because they want to combine A&E with another specialty or with a post such as Postgraduate Dean or a role within a medical school. To this is added the problems of staffing departments following the full implementation of the European Working Time Directive and new ideas about managing the hospital at night by a team based in A&E.

As I write this in 2004, there seems to be a desire by hospitals to expand consultant numbers with many new posts being advertised and without sufficient trained doctors to fill them. Expansion is not even. Many prefer a department with five or six consultants for a more civilised rota. Well-staffed departments are becoming better staffed but poorly staffed departments are not attracting applicants and may even lose consultants who apply for posts in larger departments.

There was a massive increase in new training posts in 2003/04.

6
A Changing Specialty

Increasing cover

A single-handed consultant cannot be present in the department all the time and so his or her responsibility is to organise the department so that it can cope in their absence. This involves much teaching and administration as described in the previous chapter. However where there is more than one consultant, they can spend much more time on clinical duties. In my own department 65 per cent of patients attend outside normal working hours and there has long been recognised the requirement for experienced doctors to be present 24 hours a day. In 1960 the Nuffield Report had said that the ideal casualty department 'should have immediately available at any one time a medical man of consultant quality ...'[1] and the BOA in 1971 had also recognised the need for an experienced doctor on duty 24 hours per day.[2] The possibility of consultants working shifts had been discussed following the Short Report in 1982 when there was much (but not total) opposition (see Chapter 5). It was, however, completely unrealistic with the number of consultants then available. The importance of 24-hour consultant cover (not just in A&E) was emphasised by the Royal College of Surgeons' Report on patients with major injuries in 1988[3] and in 1990 BAEM stressed the importance of providing experienced 24-hour cover wherever possible.[4]

Most departments on appointing a third consultant (and some on appointing a second) accepted that there was little point in all the consultants arriving at 9 am and going home at 5 pm and many started to work a partial shift pattern covering until, say, 8 pm or 10 pm some nights. However such departments were very much a minority with only 26 departments in England having more than two consultants by

1996.[5] One way to achieve increased consultant cover is to amalgamate A&E departments, thus providing more consultants in fewer hospitals. This idea was not new and had been recommended as far back as the 1950s with Lowden commenting on the senselessness of three hospitals in the same area all maintaining facilities for treating the same type of case[6] and both the Nuffield[1] and Platt[7] Reports recommended concentration and rationalisation which would involve closure of casualty departments. Closing a department is difficult politically and in recent years when it has happened, it has often been replaced by a nurse-led minor injuries unit. Amalgamation of departments does not always appeal to the consultants involved who find themselves working in a department and with colleagues they have not chosen. David Skinner (A&E consultant in Oxford) in an editorial in the *BMJ* in 1990 noted: 'Many consultants ... had reservations about working with a colleague of equal rank, let alone four or five consultant colleagues, but such attitudes are hard to justify. Many also disliked the idea of shift work as a consultant after years in the junior ranks. ... Yet any increase must imply a shift system to address the problem of the sickest patients being seen by the most inexperienced doctors. ... A less radical system ... would be to link groups of four or five accident and emergency departments in collaborative night cover (6 pm–9 am) to concentrate senior staff in one department.'[8]

There are other advantages of such linkage. In a city with a number of hospitals, one (usually the teaching hospital) is perceived as the 'better' hospital and so the smaller hospital may find it more difficult to attract good quality staff. If A&E departments can be linked, the SHOs can be shared and this can also have benefits for common teaching etc.

Twenty-four-hour consultant presence in A&E

The arguments for and against 24-hour consultant presence in the department have been given by Mathew Cooke and colleagues from the West Midlands in an article in the Journal of Accident and emergency Medicine.[9] Their arguments (and others) are summarised below:

The advantages of 24-hour consultant presence are:

- A&E is a 24-hour service and patients should receive the same care whatever hour they attend.
- The benefits of consultant presence are self-evident: otherwise why does an A&E consultant need training? (In fact there is little hard evidence that A&E consultants have better outcomes, largely because

it has not been looked for. There is some evidence that they have better results for major trauma.)[10]
- The presence of consultants should be able to reduce admissions. (This has been shown in other specialties but has not been looked at in A&E.)
- If the senior doctor covering the department is at home, SHOs will not phone for advice but they will ask if a consultant is there.
- It is not possible for a consultant to fully comprehend the problems of a department if they are not there for 24 hours.
- It is easier to get junior staff if they are supported and there can be shop floor teaching for registrars and SHOs 24 hours per day.
- A carefully planned shift system can be advantageous to the doctors involved.

The disadvantages are:

- Can it be justified to have a consultant on site when the workload decreases after midnight and the small number of serious cases can be dealt with by trauma team or cardiac arrest team?
- A department would need at least ten consultants.
- If a patient with an uncommon problem (e.g. drowning, snake bite, chemical contamination) attends the A&E department, a consultant will usually be involved with their management. If the number of consultants is trebled, the experience of each consultant will be less.
- Non-clinical work would suffer as management meetings, appointment committees, formal training and so on usually occur during working hours and so any consultant wanting to get involved with the management of their department or of the hospital would have to attend meetings in their own time.
- Out-of-hours work distances one from colleagues in the hospital and in the department.
- Junior staff develop an over-reliance on seniors and become reluctant to make decisions themselves. Making decisions is an important part of their professional development.
- A trainee must have some time that he is in charge.
- A single-handed consultant has his finger on pulse of the department and develops risk limitation strategies. Multiple doctors can't have their finger on the pulse.
- The department becomes dependent on constant consultant presence and problems can arise if it does not happen (e.g. due to sickness).

- Shift work has an impact on doctors. Night and weekend work mean missing out on family life.
- If A&E were the only specialty providing 24-hour cover, it would be difficult to recruit.[11]

My belief is that most A&E consultants accept that patient care would be improved by their presence throughout the 24 hours but if they did this, they would be unable to do other aspects of their job. They also do not want to practise differently from their colleagues in other specialties. In particular they do not wish to work antisocial hours for normal pay when colleagues in other specialties can earn enhanced pay from the NHS by doing additional work, out-of-hours, to clear waiting lists.

The experiment of consultants working 24 hours was first tried in Stoke-on-Trent as part of the Stoke trauma centre experiment. This was set up as a trial after the Royal College of Surgeons Report on patients with major injuries in 1988.[3] The experiment has been described elsewhere.[12] Four A&E consultants and two consultant anaesthetists provided 24-hour cover specifically for trauma but also provided cover for other emergencies. They worked in the department until it became quiet (usually the small hours of the morning) and then slept in the hospital. There were many advantages to the system but there were too few doctors involved and they did not get good support from the rest of the hospital.[13] The experiment was not a success and the conclusion was that 'any reduction in mortality from regionalising major trauma care in shire areas of England would probably be modest compared with reports from the United States'.[12] This will not be the end of the debate on major trauma centres[14] but the study proved the end of 24-hour resident consultant cover in Stoke.

The Royal London Hospital tried to implement 24-hour cover in 1995 by forming the East London A&E Consortium. This involved linkage between its own A&E department (with one consultant and a consultant in primary care) and that of the Homerton Hospital (with two consultants). There were also two registrars between the two hospitals. Twenty-four-hour cover with so few doctors proved impossible to sustain but with more consultants appointed in 1997 and with more registrars, 24-hour consultant cover was provided Monday to Friday together with full 24-hour middle grade cover. These consultants provide cover for both hospitals. More recently Newham General Hospital has joined the Consortium. By 2002 there were nine consultants in the consortium but to provide full 24-hour cover, seven days a week and to provide daytime cover for the other hospitals will require at least 14 consultants.[15]

Other hospitals also combine for an out-of-hours rota but without being on site, for example, Gloucester and Cheltenham[16] but this is usually to turn a one-in-two rota into a more civilised one-in-four or one-in-five rota.

Twenty-four-hour consultant cover was also tried at City Hospital, Birmingham in 1993 with five consultants but this proved unsustainable. After about nine months the rota changed so that there was either a consultant or a senior registrar resident at night and after a further year, it was dropped.[17]

In 1996 Epson Hospital advertised for four A&E consultants to work round the clock in shifts of 8 or 12 hours with three SHOs and nurse practitioners. These consultants would have been subordinate to a consultant working a normal contract with administrative and teaching duties. This caused much concern as they would have been consultants in name only. There were no suitable applicants.[18]

Twenty-four-hour consultant cover for most departments is not a possibility for the foreseeable future but the BAEM and FAEM in a workforce planning document in 2001 have recommended that the specialty should aim for 12-hour consultant presence seven days per week in departments seeing more than 40,000 new patients per year. This can only be achieved over ten years and will require 1350 consultants.[19]

The role of a consultant

Twenty-four-hour cover in A&E intends the consultant to be present to treat the sickest patients but there is pressure for all medical care to be provided by consultants. The Short Report[20] (discussed in Chapter 5) recommended that 'patients are entitled to be treated by trained specialists (i.e. consultants) ...' Since then 'the NHS Reforms, the New Deal, and the introduction of "Calman" training all envisage a change in the pattern of delivery of care to a service in which the majority of hospital medical care is provided or directly supervised by consultants and other career grade doctors.'[21] This has been repeated in the NHS Plan: 'The second option is to make hospital care a consultant delivered service ... It is this option that both the professions and the Government support in principle.'[22]

This does not usually find general acceptance in A&E where the view is often that, even if sufficient numbers of consultants existed to treat every A&E attender (of which there is no prospect), consultants should concentrate on the sicker patients and supervising junior staff. An article by a young consultant in 1999 argued: 'Consultants should spend a significant proportion of their working week on the shop floor, actively supervising SHOs and seeing patients as necessary. It goes without saying that they should be actively involved with or supervising the resuscitation of all seriously ill or traumatised patients in the department.'[11] Very few want to shoulder the entire workload: 'Queue shifting ... can be soul destroying and may lead to consultant burnout. Most doctors did not go into A&E medicine to spend most of their clinical time seeing sprained ankles and other minor conditions.'[11] David Yates has argued that 15 per cent of patients are sick and up to 40 per cent of patients attend inappropriately, and so up to half are appropriate but do not have life-threatening problems. 'These patients do not need consultant care. Indeed experience in North America suggests that such provision is very expensive and leads to "burnout" and dissatisfaction, with many disillusioned doctors leaving the specialty in their 40s. Careful assessment by senior house officers and registrars working to agreed protocols within a framework of multidisciplinary support is more efficient and provides invaluable experience for the trainee. ... Consultants should concentrate on ... the provision of a consultant based service for the 15 per cent who are critically ill and injured. They cannot do everything, and their patients would not expect them to.'[23]

I agree that no consultant wants to spend most of his working week seeing minor injuries but doctors would not enter A&E if they did not enjoy dealing with the problems that make up 70 per cent of the workload. It must be remembered that a minor injury is one that affects somebody else and that the minor side of the department provides diagnostic difficulties in the same way that the major side does. It also provides most of the complaints against the specialty. I believe that most A&E consultants enjoy the variety of work that the specialty provides and are happy spending a proportion of their time providing a good service for patients with minor injuries and teaching recently qualified doctors to become competent in this area.

There is evidence of greater consultant clinical involvement with increasing consultant numbers. Between 1989 and 1997 the proportion of seriously injured patients seen in emergency departments first by a SpR or consultant increased from 32 per cent to 60 per cent. This was associated in a reduction in mortality.[24]

The general view in the specialty is that 'a move towards more care being delivered by middle grade doctors (registrar/non consultant career grade) and consultants is required as a matter of urgency' with more experienced doctors seeing the sick and the high-risk. 'In addition, experienced medical staff could "sign off" the notes of all patients attending the department to ensure that high standards of care are maintained. Middle grade cover would be 24 hour on site but overnight consultant presence in departments is not proposed.'[25]

What does a consultant do?

It would be very interesting to know exactly what A&E consultants actually do in practice but the evidence is very limited and I am only aware of two studies. One looked at the workload of all doctors in 1990 as part of the evidence to be submitted to the doctors pay review body. Table 6.1 shows the workload of eight A&E consultants compared to the medical profession as a whole.[26] A&E consultants worked more hours than other specialists (54.9 hours per week compared to 49.2 hours), but the other striking difference is the greater time spent on management. This probably reflects the very large number of patients attending

Table 6.1 Workload of A&E consultants compared to all consultants 1990

	Hours of work	
	A&E consultants	All consultants
Clinical work	33.8	35.4
Teaching including preparation	3.8	2.2
Management etc.	8	4.9
Continuing education	1.2	2
Research and publications	1.4	1.2
Category two work	2.3	0.6
Domiciliary visits	0.2	0.7
Other work for NHS	1.5	0.2
Emergency recall	3	2
Clinical duties	Percentage	
Outpatients	51	24
Inpatients, ward rounds etc.	15	18
Theatres	9	21
Laboratories and department	—	13
Clinical, non hospital	2	2
Administration (patient care)	12	13
Travelling between sites	2	5
Medical audit	9	2
Others		1

Source: Reference 26.

Table 6.2 Hours worked by nine A&E consultants 1998[27]

	Mean	Range (%)
Hours per day worked	9	8–12
Percentage of time worked out-of-hours	14	2–27
Percentage of total time spent on different tasks		
Clinical	30	15–48
Administration	18	12–28 (28 was a single-handed clinical director)
Staffing	3	0–10
Teaching	21	13–38
Meetings	14	5–32
Research	12	1–32 (32 was a senior lecturer)
Reports	2	0–6

A&E departments generating a lot of pathology results and X-ray reports to be followed up together with other paperwork including complaints. A&E departments also link with a large number of other departments and specialties both within and outside the hospital.

The other study is of nine consultants working in departments with between one and seven consultants. Two were part time and seven worked late shifts in the department. The study was performed in 1998.[27] The hours worked are shown in Table 6.2.

Specialisation

It has been noted that before the NHS many casualty departments were linked with a minor operations service and this continued in many departments. (My own department continued with elective minor operations under local anaesthetic until 2001.) Most of the original consultants had a surgical background and many had come into A&E with an interest in trauma surgery. Many of the early consultants maintained an interest in surgery and operated but this tended to die out as workload in A&E increased and the number of consultant surgeons also increased. Changes in training and medico-legal concerns mean that the modern generation of A&E consultants, even those with a surgical higher qualification, are no longer qualified to operate, even if they wanted to. There had been a tendency to regard senior doctors in the specialty as being either surgeons or physicians working in A&E but most of the early senior registrars regarded themselves as A&E specialists, making no distinction between those who had entered the specialty with a medical or surgical higher qualification. Most had interests within A&E, but tended not to advertise these or regard their interest as a subspecialisation. Multi-consultant departments meant that specialisation could start again.

In the mid-1990s a few consultants in A&E with an interest in primary care were appointed. These all came from a primary care background rather than A&E and are discussed in Chapter 11.

Head injuries

In the UK, neurosurgeons have been few in number and have been concentrated in a small number of regional centres. Even in those centres, neurosurgeons have mainly only looked after patients with more severe head injuries such as those requiring surgical treatment or ventilation. In the past, most head injured patients have been looked after by general surgeons or (occasionally) orthopaedic surgeons but this has never fitted well with their other work. Head injured patients all pass through the A&E department and in 1999 about a third of A&E departments looked after minor head injuries in short-stay observation wards (about half of larger departments did).[28] With increasing specialisation, general and orthopaedic surgeons have increasingly felt unqualified and reluctant to take on the care of these patients and in 1999 a report from the Royal College of Surgeons recommended that 'those adult patients with minor head injuries who need a period of observation would best be cared for under the auspices of the accident and emergency department.'[29] Many within A&E feel that this is appropriate and about two thirds of A&E consultants are prepared to take on this responsibility if given sufficient resources[28] but there is unhappiness that the specialty seems to increasingly pick up the patients which no other specialty wants.

Paediatric A&E

Accident and emergency departments within paediatric hospitals have consultants in charge but paediatric A&E was slow in developing. Dr Cynthia Illingworth in the Sheffield Children's Hospital had been one of the first A&E consultants appointed in 1972 but these posts were slow in taking off. By 1985, Jackson, a consultant paediatrician said in a *BMJ*-editorial that there were only three paediatric A&E consultants (in Sheffield, Liverpool and Dublin) though six children's hospitals had consultant paediatricians working sessions in A&E.[30] The majority of children, however, attend A&E departments in District General Hospitals (DGH) and make up about a quarter of the A&E workload. Seven DGHs had regular sessional commitment by consultant paediatricians, three in children's A&E departments alongside adult A&E and four in general A&E departments.[30] Jackson recommended '… a consultant paediatrician being appointed to share the responsibility for the general arrangements for children and to be the contact point between the accident and emergency department and the paediatric unit'. This was not universally popular by those who were happy to liaise but worried about other specialties interfering in A&E.[31]

The SAC approved the first SR post in paediatric A&E in 1987[32] and the following year it was recommended that paediatric A&E medicine should develop as a specialty with the aim of establishing posts in paediatric A&E in both children's hospitals and in general A&E departments where many children are seen.[33] By the late 1990s there were four or five training programmes for paediatric A&E with a number of trainees. These could either come from the ranks of paediatric or A&E SpRs and the syllabuses for each had been laid down. It was recommended that all trusts with A&E departments treating emergencies in children should appoint a consultant with paediatric A&E experience and that departments seeing more than 18,000 children (approximately 20) should appoint a consultant with paediatric A&E training within five years.[34]

Pre-hospital care

A number of the early of A&E consultants (including John Collins in Derby, Keith Little in Edinburgh and Roger Snook in Bath) ran accident flying squads to attend to trapped victims at accidents and this work often continues but it became less necessary when paramedics were trained to do advanced procedures and to give drugs. Over time, the fire brigade have become better equipped for freeing trapped casualties and so this is a less common occurrence. Many consultants have also been very involved with ambulance training. There was further scope for involvement with pre-hospital care in the 1990s with many ambulance services appointing medical directors and advisors. The first of these was a full-time post with the Scottish Ambulance Service in 1992, filled by Andrew Marsden, an A&E consultant in Wakefield[35] but subsequent posts have mostly been part-time. Many (but not all) of these have also been filled by A&E consultants including Fiona Moore (London AS),

Gillian Bryce (Westcountry AS) and Jim Wardrope (South Yorkshire Metropolitan AS). An expansion of the 'flying squad' role came with the Helicopter Emergency Service based at the Royal London Hospital and the appointment of two A&E consultants (Timothy Coats and Gareth Davies) in 1997 between the A&E department and the Helicopter Emergency Service.[15] A number of A&E doctors will declare a special interest in pre-hospital care but it is not a recognised subspecialty and there is no specific training for it at present. Many A&E doctors use their resuscitation skills in their own time by providing pre-hospital care at motor racing, equestrian and similar events.

Sports medicine

A&E departments see a lot of patients with soft tissue injuries, many of which are caused by sport. Many departments run follow-up clinics for these injuries and many A&E doctors have a special interest in these. By extension, many are also interested in the more chronic overuse injuries of sportsmen and sportswomen and provide care for sporting teams. At least one A&E consultant (Michael Allen of Leicester) gave up A&E to concentrate full time on sports medicine.

Intensive care

The sickest A&E patients usually leave the department to go to the operating theatre or intensive care unit (ICU) and there has always been a close interface between A&E and ICU and anaesthesia. Some hospitals have appointed anaesthetists with sessions in the A&E department, for example, David Edbrooke was appointed with four sessions in A&E in Sheffield in 1982 and anaesthetists were appointed in Stoke as discussed above but these posts were to provide anaesthesia and resuscitation in A&E rather than to provide a fuller A&E service. There are a few A&E consultants with an anaesthetic higher qualification and in recent years there has been a tendency for more invasive monitoring and aggressive therapy to be commenced early in A&E and for the resuscitation room facilities to match those of the ICU. Anaesthetics and intensive care is an essential part of A&E SpR training and, increasingly, A&E doctors are using skills previously considered the province of the anaesthetist or intensivist. Thus in a survey of 97 hospitals seeing more that 50,000 new patients per year, rapid sequence induction of anaesthesia is done by A&E staff in 31 per cent.[36] Some A&E consultants are trained in both A&E and ICU. The first was probably William Tullett at the Western

Infirmary, Glasgow, and there are now five with a number of trainees interested in expanding their horizons in this direction and one A&E training programme designed to produce consultants with a dual CCST in A&E and intensive care.[37]

A&E and medicine

While the majority of patients attending A&E departments have suffered an injury, there is no doubt that the sickest patients are medical. Chapter 2 demonstrated that the data in the Platt Report on which it was recommended that A&E departments be led by orthopaedic surgeons, show that the biggest proportion of admitted patients were medical. In the early 1970s the patients going through the resuscitation room were also predominantly medical.[38] Other work showed that most of the deaths within A&E or within five days of admission from A&E were medical.[39,40] Since then the medical workload of most departments has increased and this is associated with a decrease in the amount of major trauma. This is due to the reduction in road accidents and also the reduction in industrial accidents which is, in part due to improved Health and Safety legislation but also due to a much smaller industrial base in the UK. Not only have the number of medical attendances increased but so has the proportion and the number admitted.[41-43] The reasons for this are outside the scope of this book but it has put enormous pressures onto A&E departments as medical patients are more time consuming to deal with than most trauma patients and with bed shortages, medical patients may wait for admission on trolleys for many hours in A&E. One specific problem is patients with myocardial infarction. Thrombolysis has revolutionised the treatment of this condition but, to be effective, it needs to be done rapidly. A&E departments have demonstrated that they can do this as effectively as, or better than, physicians or coronary care units: 'The shortest door-to-needle time occurred when thrombolysis was administered in casualty.'[44] Without doubt the way that A&E departments have improved the treatment of this (and other conditions) has improved the status of A&E departments considerably but has increased the workload in so doing.

The recognition that much of the workload of an A&E department is medical has increased the proportion of doctors in the specialty with a medical higher qualification. In the 1970s the vast majority of A&E doctors had FRCS but by 2000 it was said that 20 per cent of A&E consultants and 30 per cent of SpRs had MRCP.[45] Of those with a surgical qualification, probably the majority now have the Edinburgh A&E FRCS or MRCS rather than a surgical FRCS or MRCS and most of them will have obtained the part 1 MRCP as a primary qualification to allow them to sit the exam.

Another pressure is trying to reduce hospital admissions while, at the same time, preventing medico-legal claims for misdiagnosis. This has led to streamlining the investigation and management of deep vein thrombosis and, in patients with chest pain, developing strategies for ruling out myocardial infarction sometimes with the use of a chest pain investigation unit.[46] These (and other protocols for ruling out specific problems) are all suitable to be done within A&E or in a clinical decision unit attached to A&E and have been adopted by a number of departments.

At the same time as this increased medical workload, general physicians are becoming increasingly specialised and more reluctant to be 'on take' for general medical admissions.[47] One view was that 'it is unfair to specialist consultants and to their patients to continue to expect the former to practise general medicine.'[48] In some places acute physicians have been appointed just to look after the acute admissions unit. (Initially some called themselves emergency physicians which caused confusion, as emergency medicine is the name given to A&E in the USA and Australasia and doctors who practice in those countries call themselves emergency physicians, as do some A&E doctors in the UK.) The appointment of acute physicians was, initially, discouraged by the Royal Colleges of Physicians but in 2000 was accepted as a subspecialty with a potential role for dual training in A&E and general internal medicine.[45] A further report from the Royal College of Physicians of London in 2004 takes this much further and recommends that by 2008, hospitals should have three acute physicians (more in larger hospitals) with a commitment to A&E departments. They also see a role for suitably trained A&E consultants to practise acute medicine.[49]

This expansion of A&E is exciting but it is not without potential problems and it is important that expansion is planned and that it improves patient care so that A&E is not just pushed into looking after the problems which other departments do not want. Any expansion in the role of A&E must be associated with appropriate training and increased resources. In particular if A&E consultants are to do intensive care or run medical admissions units, there need to be more A&E consultants to cover the A&E department or there will be a new generation of absentee landlords.

7
Academic A&E, the Faculty and Changes of Name

For a long time A&E was regarded as a Cinderella specialty and until the early 1970s it was not recognised as a specialty at all. Following the appointment of the early consultants it was commonly regarded as a surgical subspecialty or, even worse in the eyes of its consultants, an orthopaedic subspecialty. And yet it was recognised that it dealt with medical emergencies. The first way to define a specialty is to develop a training programme and this was done in the late 1970s as described in Chapter 4. A specialty-specific exam is also useful and was achieved in 1982 but it took a long time to become fully accepted as the equivalent of the other higher diplomas.

Once the specialty was accepted, it became necessary to improve its status. Most human groups seek to improve their status and image and medicine is no different. Improved status does, of course, help a doctor's self-esteem and assist his own career advancement but there are other reasons. High status specialties find it easier to recruit new members and easier to obtain resources for their departments. There will also be advantages in hospital politics when negotiating with management and colleagues. Providing a good clinical service is obviously essential to achieve this but other things are needed. One of these is academic development.

Academic A&E

Academic medicine is involved with teaching and research. Medical students have gained experience in casualty departments since long before the NHS though organised teaching by A&E staff was less common and varied between medical schools. Properly organised teaching requires additional staffing. Teaching has probably reflected the workload of A&E departments with much of the early teaching concentrating on wound care and fractures though consultants actively involved with A&E recognised other roles. Patrick Clarkson recognised casualty as a place for learning clinical responsibility for decision-making, maintaining standards under pressure, cooperation with others and accurate documentation.[1] All A&E consultants recognise the value of student attachments to A&E and David Wilson summed up a common belief that 'any doctor will be ill-clothed who has not had this experience [an attachment to A&E] and any curriculum is slightly threadbare that does not include it'.[2] A&E also needs to create a good impression on medical students so that they will return after two or three years to work as SHOs.

Research and audit is essential for the development of any specialty and this, too, was recognised by the early pioneers of casualty and A&E. In 1955 Lowden wrote: 'The casualty officer will do well if, in his mind, he questions every one of the accepted processes from time to time, and enquires whether they have a rational basis or a traditional one. It would be impossible to carry out controlled investigations into every query which would arise, and chaos would result if more than a minute fraction of his questions were put to the test. Every officer can, however, investigate a few, and the sum total of such investigations would be valuable.'[3] It has been noted in Chapter 3 that Maurice Ellis, too, undertook research.[4,5] One of the aims of the newly formed CSA was 'to promote interest in and by means of investigation to further the knowledge of accident and emergency work in the hospital service' and at the first meeting they decided to investigate the misuse of casualty departments; the time spent by casualty officers in specific duties and the ultimate source and reason for attendance.[6] Research was, however, a low priority when faced with more than enough clinical work and political battles to fight. One or two of the early consultants (e.g. Keith Little in Chester and, later, Edinburgh and Roger Snook in Bath) had done research leading to an MD but there was no organised academic activity by A&E doctors. However A&E was significantly influenced by the research of others, for example, head injury research by Prof. Jennett and colleagues in Glasgow and the care of patients with myocardial infarction by Prof. Pantridge in Belfast and Douglas Chamberlain in Brighton.

There can be no argument that the centre of academic A&E in the UK is Manchester. Prof. Irving's interest in trauma has been mentioned in Chapter 4 and this was a major factor leading to the decision by the Medical Research Council to locate its new Trauma Unit in Manchester in 1977. The Trauma Unit was led by Prof. B. Stoner who worked on the metabolic and haemodynamic response to trauma. When Prof. Stoner retired, the leadership passed to Prof. Roderick Little[7] who was later the first chairman of the Faculty of A&E Medicine's Research Committee. In 1977, in an article in 'Resuscitation', Prof. Irving emphasised the importance of academic development in A&E (which he called Emergency Care) and he felt that the best way for the new specialty to achieve this was to look for one or two academics within the ranks of senior registrars and to allow them to train in established academic departments and research units.[8] They could then become senior lecturers within friendly existing academic departments before persuading universities to develop chairs in A&E. This is almost exactly what happened. Prof. Irving appointed David Yates as tutor and honorary senior registrar in A&E in 1976 and on completion of his training, he was appointed as senior lecturer in A&E in Manchester and consultant at Hope Hospital Salford. In 1990 he was appointed to the first chair of emergency medicine in the UK in Manchester. He has worked very closely with the MRC Trauma Unit and has trained a number of academic senior registrars and specialist registrars who have gone on to obtain senior lecturer posts and chairs. His unit also set up the UK Major Trauma Outcome Study (MTOS) which later became the Trauma Audit and Research Network (TARN) which audits the outcome of major trauma in hospitals in England and Wales.

In 1992 Graham Page, an A&E consultant in Aberdeen, was awarded an honorary chair in emergency medicine at Robert Gordon University for his work in offshore medicine[9] and in 1995 Anthony Redmond who had trained in Manchester, was appointed to a chair at Keele University. Subsequently further A&E doctors have been appointed to senior lectureships and chairs and others have been attracted from outside the specialty to chairs in A&E (John Henry at St Mary's Hospital, London and James Ryan at University College Hospital). There are also a number of academic and research registrar posts, many of which are based on Manchester. Good research does not necessarily require academic posts as has been shown by much work on resuscitation, cardiac arrests and pre-hospital care that has emerged from Edinburgh under the supervision of Colin Robertson and Keith Little.

A number of other A&E departments have sought to make senior academic appointments and some have met with resistance from academics in more established specialties. Even when hospitals and universities have been supportive, many have failed to appoint due to lack of A&E doctors with appropriate academic training. Gradually, however, suitably qualified doctors are becoming available but funding both the posts and research remains a problem. While it is recognised that there is a great need for more research to provide the evidence on which to base A&E practice, there are major problems for academic medicine in the early part of the twenty-first century and on David Yates' retirement, there are no plans to replace him.

More recently the Faculty of Accident and Emergency Medicine has awarded (unfunded) Professorships to Kevin Mackway-Jones (Manchester Royal Infirmary) and Robin Touquet (St Mary's) as an honour and to assist them in furthering their own research.

A multi-centre study organised by William Rutherford looking at the pattern of injury in road accidents before and after the introduction of compulsory seat belts was a model of what A&E could achieve by departments collaborating.[10]

In 1983 David Yates, Rod Little, Keith Little and William Rutherford convened a meeting which founded the Emergency Medicine Research Society (EMRS) to 'further research and a common interest in accident and emergency medicine and related topics'.[11] (Rod Little's name does not appear in the reference quoted presumably because he wrote it.) The EMRS held successful annual meetings for about ten years. A number of these were held as joint meetings with the US Society for Academic Emergency Medicine.

The Faculty

When the SAC was established in 1975 it was answerable to both the Medical and Surgical Joint Higher Training Committees (JCHMT and JCHST) which complicated decision-making. On appointments advisory committees (AACs) for A&E consultants there were representatives of the Colleges of both Physicians and Surgeons but neither had to be an A&E consultant. In practice the Royal College of Surgeons usually chose an A&E consultant to represent it but the Royal College of Physicians did not do so routinely until the late 1980s or 1990. A&E was thought of as a small specialty: it was in consultant numbers, but in workload, it was bigger than almost any other specialty. The medical world was (and is) full of people who knew how to run A&E departments much better than those who worked in them and the CSA regularly used to complain that specialist associations would write guidelines for management of A&E problems without any consultation with those who do it every day.[12-14] The DHSS advisor on A&E was an orthopaedic surgeon[15] until Howard Baderman was appointed advisor in 1990.[16] When the GMC started to register completion of higher specialist training, A&E doctors with an FRCS as a higher qualification were registered as 'T(S)' (trained in surgery), whereas identically trained doctors who happened to have an MRCP were registered as 'T(M)' (trained in medicine).[17] The Royal College of Surgeons of Edinburgh had always been very supportive of A&E but the specialty would have preferred to be in control of the only higher diploma in A&E. Since its inception, the specialty has been trying to assert its independence. A Royal College was probably too ambitious a goal but a Faculty of Accident and Emergency Medicine would do this and would also improve the status of the specialty.

The first discussion about forming a Faculty was in 1974 when, during planning for SR training programmes, two consultants in the North West Region were advised by the Postgraduate Deans of Manchester and Liverpool that 'informal discussions should be held with the College of General Practitioners to explore the possibility of establishing an independent faculty of that College. They suggest that the Royal College of General Practitioners is a highly appropriate body with which to be associated'. The Deans advised that it was unlikely that the Royal College of Surgeons would accept a casualty faculty and the BOAs involvement might prove a hindrance. Drs Malcolm Hall (Preston) and Gordon Laing (Salford) disagreed with the view that a Faculty should be of the College of General Practitioners but said that their views were founded less on logic than on tradition and emotion.[18]

Linking with the Royal College of General Practitioners was also not acceptable to the 1975 CSA AGM[19] who believed that it would be more appropriate to affiliate with the Royal College of Surgeons. In 1976 the Chief Medical Officer advised the CSA not to pursue the idea of a Faculty at that time[20] but in 1977 some members felt that if becoming a Faculty proved 'impossible we should establish ourselves as an independent College – A "College of Emergency Medicine and Surgery" '.[21] The Lewin Report[22] mentioned the idea of a Faculty, only to dismiss it. Howard Baderman looked into the possibility but told the 1980 AGM that he was forced to conclude that there was no prospect of establishing a Faculty within any of the Royal Colleges.[23]

It has already been noted (Chapter 4) that in the early 1980s the Royal College of Surgeons of Edinburgh started two specialty exams (often described as exit exams as they signify the end of training). By the late 1980s, all the Royal Colleges of Surgeons were considering changing their exam structure. This led, in 1988, to David Williams, President of the CSA, setting up a small group (Future Strategies Group) to consider the future of examinations in A&E and the specialty itself. Members were the officers of the CSA (David Williams, Maj. Gen. Norman Kirby, Gautam Bodiwala, and John Thurston), and other senior members of the specialty (Howard Baderman, Keith Little, Stephen Miles, Colin Robertson, David Wilson and Prof. David Yates). Their consensus was that steps should be taken to explore the possible formation of a Faculty which would ideally be independent of any College.[24]

By the following year, the Colleges of Surgeons had decided that each surgical specialty would have an exit exam and this would be an intercollegiate exam run by all four surgical Royal Colleges and that the FRCS would not be awarded until this exam had been passed. It was clear that changes would have to be made to the A&E exam to fit in. An intercollegiate board was set up to look at and plan the new exam for A&E[25] and there was widespread agreement to set up an exit exam for A&E[26] similar to the other surgical exit exams. Without a Faculty or College, the specialty would be a very junior partner in any intercollegiate exam and there was encouragement for the formation of a Faculty by the Royal College of Surgeons of England. To become a College would have necessitated purchasing or leasing a building and that would have been too expensive. The Future Strategies Group did not want to become a Faculty of the Royal College of Surgeons of England alone as A&E has spent a long time persuading people that it was not a surgical specialty. They wanted involvement of the Royal College of Physicians and did not want to exclude the Royal College of Surgeons of Edinburgh which had been a strong supporter of A&E. Their recommendation was an intercollegiate Faculty.[26] The Royal College of Physicians of London was opposed on principle to the establishment of further faculties[27] and the Royal Colleges of Physicians of Edinburgh and of Physicians and Surgeons of Glasgow expressed similar views.

The first meeting of the provisional Intercollegiate Board to plan the exam was held in May 1991. The members were David Williams, Maj. Gen. Norman Kirby and Howard Baderman (all representing BAEM), Prof. David Yates (RCSEng), Ian Anderson (RCP&SGlas), Keith Little (RCSEd), Colin Robertson (RCPEd), Peter O'Connor (RCPIreland), E. Beck (RCPLond), Prof. N. O'Higgins (RCSIreland) and Peter Baskett (Royal College of Anaesthetists (RCA)). It was reassuring and a sign of the progress that the specialty had made, that five of the eight Colleges had asked A&E consultants to represent them. After the establishment of the provisional Intercollegiate Board, the Working Group set up to establish the exam had continued but had changed its name to the 'Working Group to consider the establishment of a Faculty of Accident and

Emergency Medicine'. This met first in March 1991 and considered draft standing orders for the Faculty and a discussion paper concerning the relationship between the Faculty and BAEM.[28] Essentially the Faculty would be responsible for all training, education and academic matters and BAEM would be responsible for service matters. There is, of course, significant overlap and the president of BAEM sits on the Faculty Board and vice versa.

The existence of the Faculty of Accident and Emergency Medicine (FAEM) owes its existence largely to the diplomatic skills of David Williams who managed finally to win support from the Royal Colleges for its establishment. By April 1992, there was sufficient support to form a Faculty but there was still uncertainty as to whether the Royal College of Physicians of London wished to be associated with it and it was thought that it might be necessary to form the Faculty without that College's involvement but with the possibility of it becoming involved at a later date.[29,30] Eventually the College agreed and a Steering Group was set up to establish the FAEM. This consisted of David Williams (chairman), Gautam Bodiwala, Maj. Gen. Norman Kirby, Stephen Miles and Prof. David Yates, all representing BAEM, Prof. Norman Browse (RCSEng), Dr David Pyke (RCPLond), Dr J. Nimmo (RCPEd), Mr A. Dean (RCSEd), Mr T. Hide (RCP&SGlas), Dr J. Stoddard (RCA) and Dr Keith Little (Chairman SAC). The Irish Colleges were not represented in the Faculty but continued to be represented on the Intercollegiate Board considering the exam. The Steering Group was later replaced by a Working Party with largely the same membership.

An advertisement in the *BMJ* and circulars to the BAEM membership invited consultants and associate specialists in substantive posts to apply to become Founding Fellows and the Faculty was inaugurated at a ceremony at the Royal College of Surgeons on 2 November 1993 with 317 Founding Fellows (though two subsequently withdrew). The Faculty shared an office with BAEM in the Royal College of Surgeons. The founding Fellows elected the first officers and members of the Faculty Board who were announced at the Faculty's first General Meeting on 14 December 1993 at a conference on 'A&E Medicine and the Health of the Nation' at the Royal College of Physicians. Not surprisingly Dr David Williams was unopposed and became President. The full list of members of the first Faculty Board is shown in the appendix (Appendix B).

Once the Faculty was formed, the Intercollegiate Board became the Examination Committee of the Faculty (later becoming the Education and Examination Committee) and the Faculty took on the responsibility for nominating independent doctors to serve on complaints procedures. As from 1 January 1995 it took over the responsibility for nominating advisors for AACs for A&E consultants.[31] Another early task was to establish its identity further by acquiring a coat of arms. This is described in an article by John Thurston, the founding Registrar.[32] There were other tasks such as establishing systems for monitoring continuing medical education.

It was recommended that the EMRS be invited to join the Faculty with the EMRS funds being ring-fenced for research and this subsequently happened. Rod Little became the first (and, to date, the only) non-medically qualified Fellow of the Faculty and later became the Chairman of the Faculty's research committee. A prize for the best research paper presented by a trainee at the Faculty's annual conference was named the EMRS prize.

The exam set up by the Faculty was the FFAEM exam to be taken at the end of training and it was essential to pass the exam to complete training and become a consultant. The first exam was on 3 October 1996 when there were four candidates, all of whom passed.

Despite the formation of FAEM, training in the specialty was still the responsibility of the SAC. Although the FAEM had representation on the SAC, it was still answerable to both the JCHMT and the JCHST. It was considered vital that the specialty take on the responsibility for its own training and in 1994 the Faculty Board agreed that it wanted to take over the function of the SAC.[31] Again Dr Williams' (and others') diplomatic skills were needed as not only was there diversity of opinion in the Colleges as to whether this could happen but reluctance within the Department of Health to give money for the administration of training to other than Colleges. However the JCHTA&E came into being on 1 January 1998. This was the responsibility of the FAEM but it was administered on its behalf by the JCHST.

Unfortunately the setting up of the JCHTA&E still did not give the specialty complete independence. In the early 1990s the European Commission had expressed concerns that the system in place in the UK for the mutual recognition of specialist medical qualifications did not fully comply with the 1975 European Medical Directive. A working party was set up under the chairmanship of the Chief Medical Officer, Sir Kenneth Calman which produced what became known as the Calman Report.[33] It produced many recommendations, some of which are discussed in other chapters, but one of these was that there should be a body legally responsible for safeguarding standards of training and ensuring that training requirements adhered to the European Medical Directive. It would issue certificates indicating the completion of specialist training. This body, set up in 1977, was the Specialist Training

Authority of the Medical Royal Colleges (STA). Its membership included representatives of all the medical Royal Colleges and the Faculties of Public Health Medicine and Occupational Medicine.[34] These were the Faculties represented on the Conference (later Academy) of Medical Royal Colleges and Faculties from which the FAEM had been excluded (the only Faculty with training responsibility which had been so excluded). Because of its legal responsibility, the STA had to approve training syllabuses and the regulations of each College for recommending a Certificate of Completion of Specialist Training (CCST) and regulations for admitting others (e.g. those trained outside the EEC) to the specialist register. The STA would not communicate directly with the FAEM and all communication had to be done via the JCHST which was represented on the STA. Not only did this cause difficulties in communication and delays[35] but it caused confusion in that the JCHTA&E was perceived to be no different from the SACs in the surgical specialties which report to the JCHST, particularly as the administration of the JCHTA&E was done within the JCHST office.[36,37]

In 2001 the Faculty was admitted to the Academy of Medical Royal Colleges and in 2002 the administration of the JCHTA&E was moved to the FAEM office but the JCHTA&E still has to communicate with the STA via an intermediary.

There are many within the Faculty now who would very much like it to become a College.

What's in a name?

A&E departments came into being in 1962 with the Platt Report but they have been called many things including 'accident departments', 'emergency departments' and 40 years after Platt, they are still often referred to as 'casualty departments' despite pleas that 'the term casualty ... be banished from use in the *BMJ* in the future'.[38] The specialty has been called 'accident and emergency medicine' since its formation in 1972 and it had worked in accident and emergency departments but the association had remained the 'Casualty Surgeons Association'. The name had been criticised by some almost as soon as the association was formed[39] and changing its name had first been seriously discussed at a committee meeting in 1972 but further discussion was deferred.[40] Motions for changing the name were submitted to the AGMs in 1980 and 1986 and although there was a majority for change at the later date, the motion failed as a constitutional change required a two-thirds majority. Feelings ran very high on both sides. One view was expressed by the editor of the *British Journal of Accident and Emergency Medicine*: 'I wish to make it clear that I intend to have no truck with the term "Casualty", an archaic handle which I consider to be on a par with "apothecaries" and "Barber surgeons". The term will be ruthlessly expunged from manuscripts before they are allowed onto the pages of this journal! The term ... sums up all that was bad about our specialty in those days [before the Platt Report in 1962]. It is high time that it was consigned to its rightful place in history. That the term in question remains part of the title of our association after a majority vote against it at the last CSA Annual General Meeting, is a quirk of constitutional democracy, for which, at present, there is no remedy.'[41] The other extreme is illustrated by the fact that when the name eventually changed to the 'British Association for Accident and Emergency Medicine' (abbreviated as 'BAEM') at the 1990 AGM, two founder and honorary life members (Edward Abson and Alec Murray) walked out and resigned. This was a great pity as Edward Abson, in particular, had had such a significant role in the formation of the specialty.

As has been noted, the specialty of A&E in the USA and Australasia is called 'emergency medicine', possibly reflecting its different origins. This name has the advantage of being short and there were pressures to change the name of the specialty in the UK. The possibility of changing the name of the CSA to 'The Society of Emergency Physicians' was first raised in 1972.[40] The Emergency Medicine Research Society was formed in 1983 and the journal, *Archives of Emergency Medicine*, was started in 1984 but these were aiming to attract a wider membership and circulation than just A&E.

Chapter 6 describes how, by the late 1990s some hospitals were starting to appoint acute physicians. In some places these were called emergency physicians, the term by which A&E doctors are known in the USA, Australia and other countries. This had the potential to lead to confusion. One consultant summed up the feelings of many when he feared that A&E might lose the right to the name 'emergency medicine'.[42] Those wanting to contribute more to the care of medical emergencies feared that this might be hindered by 'the stubborn retention of the UK-specific name "accident and emergency medicine" '.[43] An editorial in the *BMJ* also supported a change of name.[44] Along with changing the name of the specialty, those who practised it would change from being 'accident and emergency specialists' to 'emergency physicians'. Changing was not without problems. The specialty had already moved from 'casualty' to 'accident and emergency medicine' as being more descriptive of the work done and it continued to 'see and treat both accidents and emergencies'.[45] In the USA the word 'physician' covers all medical graduates including surgeons whereas in the UK the word is used just for those practising internal medicine and its subspecialties. Having spent many years trying to persuade the profession that it was not a surgical specialty it did not want to be thought of as purely a medical specialty.

A questionnaire sent to members of the specialty in late 2002 showed a large majority in favour of a change of name and the 2004 AGM, BAEM changed its name from British Association for Accident and Emergency Medicine to British Association for Emergency Medicine. Its abbreviation, conveniently, remained the same. A name change for the specialty has been approved by the STA and should happen in October 2004[46] and the name of the Faculty will undoubtedly change after that.

8
Non-consultant and Non-training-grade Doctors

This chapter discusses those A&E doctors who are neither consultants nor in the training grades since the start of consultant-led departments. The senior casualty officers could have been counted in this group but are discussed in Chapter 1. The group to be discussed are medical assistants (renamed associate specialists in 1981) and staff grade doctors (collectively called non-consultant career grade doctors – (NCCGs)) and the part time posts of clinical assistant and hospital practitioner. (It is worth noting that the term NCCG is disliked by those who hold such posts as it defines them not by what they do but rather by what they are not. An alternative is SAS doctors – staff and associate specialists. It is likely that these two grades will be combined into a single grade soon.)

The Medical Assistant (MA) grade was recommended as a replacement to the SHMO grade by the Sir Robert Platt Committee in 1961.[1] His report said there was a need 'for a grade of unlimited tenure below consultant rank and clearly distinguishable from the consultant grade. It should consist of doctors who will work as assistants and be open to doctors with two or more years' service as registrars. ...' Vacancies should be advertised and filled by competition. He felt that the MA grade 'should help with the staffing problems of casualty departments' as long as the MA was 'fully embodied in the "team"'. Sir Harry Platt in his report, agreed.[2]

The senior casualty officers transferred to this grade and there were new appointments to it. Many became consultants in A&E in the early and mid-1970s but a number did not and remained as medical assistants. The MA grade proved unpopular for two reasons. The first was its name which might suggest that the holders were not medically qualified. The second was the concept that they would only work as an assistant to a consultant and not treat patients independently. The Todd

Report in 1968 recommended that it be abolished[3] but the following year a report on the responsibilities of consultants saw that there was a need for 'those few doctors who after training have been unable to obtain vocational registration and yet who are suitable and personally apply to remain in a whole-time hospital post with limited responsibility. This is not however a grade for which there should be an establishment, and each appointment should be specifically authorised for each individual at his request'.[4]

The Lewin Report noted that at the end of November 1976 there were 76 MAs in A&E in post in England and Wales. Twenty two had a higher qualification and 54 did not. A moratorium on new posts was declared on 14 July 1977 at the same time as the moratorium on consultant posts.[5] The moratorium was eventually lifted and in 1981 the grade was renamed 'Associate Specialist'. Since then the post has continued as a post for those who do not wish to or cannot obtain consultant status. It continues to be allocated on an individual basis only. It is fair to say that

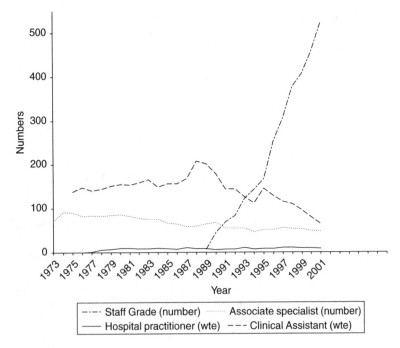

Figure 8.1 Number of associate specialists, clinical assistants, hospital practitioners and staff grades working in A&E in England 1973–2001
Source: www.doh.gov.uk/stats/history.htm

the need for the grade is accepted but there remains much dissatisfaction with it. Many associate specialists have felt that they have insufficient status and financial reward. While they may be happy working with an individual consultant this may change when that consultant leaves or retires and is replaced with a younger consultant less experienced than themselves. There is also a general unhappiness in the medical profession (particularly junior doctors) about the idea of a permanent subconsultant grade. Figure 8.1 shows how numbers have slowly decreased.

A few associate specialists continued to be appointed consultants until 1 January 1997 when it became a requirement to be on the Specialist Register to be appointed to such a post. After that date there was a transitional period for doctors who had not completed a training programme but who had equivalent training to apply for a place on the Specialist Register. Fifty A&E doctors applied and six were advised that they needed less than one year's training to be admitted to the Specialist Register.[6] At least one associate specialist did this and is now a consultant.

General practitioners working in A&E

It has been noted in Chapter 1 that GPs have worked in casualty departments since before the start of the NHS. In most cases, this will have been as a means of seeing new patients and coping with the workload. Many departments relied on GPs when they were short of SHOs. I became the first A&E consultant at the Royal Hospital, Wolverhampton in 1983 when there was a need for 26 clinical assistant sessions per week to maintain basic cover. However when I took on the post there were only two SHOs in post out of an establishment of five and so 56 clinical assistant sessions per week were being worked by about 20 different doctors, the majority of whom were GPs. Although employed to see new patients, experienced GPs would be turned to by junior hospital doctors, for advice. In some places GPs were employed in a supervisory capacity to 'give continuity between one lot of casualty officers and the next, helping to avoid the constant repetition of organisational errors'.[7] In the late 1960s and early 1970s it was considered, by some, that they should run A&E departments – this is discussed in Chapter 2. GPs are usually employed in the clinical assistant grade. A CA is a part time medical officer but the grade is not properly defined and does not appear in the terms and conditions of service. The only requirement is to have full registration. There is a defined salary but with no increments and no right to study leave. It is therefore not an attractive post. CAs are usually given one-to-three-year contracts but can be appointed indefinitely. In 1976 the Hospital Practitioner (HP) grade was introduced to make part time hospital appointments more attractive for general practitioners. HPs have a higher salary scale and better conditions of service but the grade is only open to principals in general practice who are limited to a maximum of five sessions per week. The HP grade has never been particularly popular in hospitals. Part of the reason is that they cost more than CAs but there has been a feeling that the grade was unfair as CAs who are not GP principals are unable to obtain the enhanced salary and conditions of service. Figure 8.1 shows how few hospital practitioners there are compared to clinical assistants.

GPs are used in A&E in three ways. First they may work one or two sessions per week on a regular basis to keep an interest in hospital medicine or to maintain specific skills. In the past a number of doctors went into general practice when they had failed to progress up the hospital training ladder and for some of these, hospital medicine of some sort remained their real medical interest. Not infrequently GPs are employed for, say, one afternoon per week to allow the junior staff to attend teaching sessions. On occasions GPs are recruited to fill gaps in the rota. (The wise A&E department manager knows which local GPs have overdrafts!) Last they may work sessions in A&E specifically to provide a primary care service: this is discussed in Chapter 10. Their use was recommended by the Platt Report: 'The GP can make a real contribution to the staffing of a hospital accident service. His experience...makes him particularly suited for sorting the casual attenders.'[2] One advantage of GPs was that they can 'more readily return to their general practitioners many cases which normally return to the hospital for follow-up treatment'. However there could be disadvantages as well. A report on medical staffing in 1971 noted: 'Complaints were made at one hospital, that where general practitioners took over the accident and emergency department due to shortage of hospital doctors, the number of admissions to hospital rose dramatically.'[8]

The Lewin Report considered that general practitioners could work in A&E to 'assist with the load' of minor injuries and he also felt that the presence of GPs in A&E would benefit trainee GPs working in the department.[5] The Mills Report noted the contribution of general practitioners and considered that the Hospital Practitioner Mark 11 (a grade of hospital practitioner proposed for doctors who were not principals in general practice but which was never implemented) might improve staffing in A&E and that 'more women doctors may be attracted to taking up part-time jobs within the field.'[9]

Around this time general practice was developing and becoming more self-confident and the Short Report in 1981 felt that 'most GPs would be better employed looking after their patients outside hospital...in the community' as there was still a vast unmet need for preventative medicine and health education that GPs should be devoting themselves to. However some GPs 'welcome an opportunity of linking-in with the hospital service...we gain and maintain an additional skill...' and the Report concluded: 'We believe therefore, that the development of the Hospital Practitioner grade for GPs should be fostered, and we are not impressed by arguments that this will hinder the development of general practice as a specialty in its own right.'[10]

The CSA however did not want to have to rely on GPs for the staffing of A&E departments.[11] One of the reasons for this was experience of GPs 'placing their practice commitments before their responsibility to the hospital'.[9] GPs, too, while happy to do some daytime sessions, were reluctant to work evenings, nights and weekends and so their help was of little value in providing a 24-hour service.

Nine-session clinical assistants

While the intention of the clinical assistant grade was that it should be filled by GPs, it has also been open to others. Industrial medical officers (and similar) might wish to maintain contact with the hospital in the same way as GPs and doctors with other commitments who wished to work one or two sessions only have also used this grade. Junior doctors pressed to work extra shifts often demanded employment as a CA as that paid more than a locum SHO. The maximum number of sessions as a clinical assistant used to be nine per week and some doctors were employed as nine-session CAs. This provided A&E departments with more experienced doctors who could be used as an intermediate grade. It also provided employment for those who were unable to progress with their training but who were not qualified for a post as a MA or Associate Specialist. At the time when there was a moratorium on MA posts, it was the only grade open to doctors who wanted to work in A&E but who were not able to obtain an SR post.[12] In the early 1980s an increasing number of nine-session clinical assistant posts were advertised in a number of specialties of which A&E was one of the most common.

The medical profession, and particularly junior doctors, are resistant to the idea of a permanent subconsultant grade of any sort. A service provided by such a grade costs less than a service provided by a consultant and may be seen as a cheap alternative to consultant expansion. This may harm the career prospects of trainees. Individuals within such a grade may be very competent but they have not received the training which a consultant has received and so there is the possibility that such a service will be less good than that provided by consultants.

The views of A&E doctors have differed from much of the rest of the profession on this due to the very different workload and pattern of attendance compared to other specialties. The requirement for experienced doctors within an A&E department for 24 hours a day has been long recognised and increasingly it has been accepted that every patient should either be seen by a senior doctor or for their management by a junior doctor to be supervised by a senior doctor. This requires a large expansion in experienced doctors working in A&E. The reluctance of consultants to provide the routine care of A&E patients 24 hours per day has been noted in Chapter 5 and the number of trainees must relate to the number of consultant vacancies rather than the service load. The CSA frequently recommended that a permanent subconsultant grade was relevant for A&E as did the Robert Platt Report and 'Achieving a Balance'.[13] A&E trainees have sometimes sided with the junior doctors and have opposed the idea of a subconsultant grade[14,15] and sometimes have supported the CSA.[16]

By 1984 junior doctors were becoming concerned about the number of nine-session clinical assistant posts being advertised.[17] Table 8.1 shows the number of clinical assistants in A&E in 1984. Eighty-three

Table 8.1 Clinical assistants in A&E 1984

	Non-GPs	GPs
No. of CAs	206	486
WTE	80.9	76.6
Sessions		
1 or less	67	322
2	24	81
3	20	35
4	15	19
5	10	8
6	2	5
7	4	2
8	8	6
9 or over	56	8

Source: Reference 18.

per cent of general practitioners were working two or fewer sessions per week whereas there were 56 non-GPs working nine sessions or over.[18]

Recruitment to nine-session clinical assistant posts ceased after the introduction of the Staff Grade (see below). Figure 8.1 shows the number of whole time equivalent (WTE) clinical assistants. The graph combines the GP clinical assistant, the nine-session clinical assistant and others which makes it difficult to interpret trends. However the slow rise in the mid-1980s probably represents nine-session clinical assistants. The fall after 1988 probably indicates the transfer of nine-session clinical assistants to staff grade posts. However between 1988 and 1999 there has been a fall in 112.8 WTE clinical assistants. This is more than the number of nine-session clinical assistants and so represents a falling number of general practitioners working sessions in A&E. There are probably a number of factors causing this, including better staffing of A&E departments with other grades of doctor and increased workload in general practice. It is also not financially viable for many GPs to work in a hospital as they earn less than it costs them to employ a locum to cover their practice. In 2003 a new grade was proposed of a GP with a special interest. Some of these may work in A&E or may provide emergency care in other ways.

Staff grade

A new non-training intermediate grade was recommended in 1987 in the negotiations over 'Achieving a Balance'[13] and this was later called Staff Grade. Post holders should have had a minimum of three years as an SHO. It was intended that staff grade doctors should not be used for on-call work but should work in specialties either where there was no on-call or where out-of-hours work involved continuous intensive work. A&E was specified as an appropriate area for staff grade doctors to work.[13]

Following the introduction of staff grade posts in 1988/89, no further CA appointments were allowed for more than five sessions per week. Existing nine-session CAs could keep their contracts but they were advised to apply for staff grade posts.[19] However all staff grade posts had to be advertised and so CAs applying for their post to be upgraded to staff grade risked losing their job if they were not appointed. In order that staff grade appointments did not reduce consultant expansion, posts (in all specialties) were meant to be limited to 200 per year for the first five years and 100 per year thereafter with numbers never exceeding 10 per cent of consultant numbers.

However in the mid-1990s there was a great shortage of SHOs and the ceiling on new staff grade posts in A&E was lifted (see Chapter 9). Since then, there has been increased importance given to having experienced doctors present in the department to supervise in experienced SHOs and, since 'Reforming Emergency Care'[20] in 2001, importance given to having experienced doctors seeing patients with minor injuries as a means of reaching government targets. It is probably unreasonable to ask any doctor in a career post to work more than one weekend in four and since the European Working Time Directive, time resident in the hospital is counted as work even if the doctor is asleep. To have 24-hour resident middle grade cover requires a minimum of eight doctors. These cannot all be filled with trainees and so there will undoubtedly be further expansion of this grade.

Who are staff grades?

There have been two surveys of staff grade doctors in all specialties. Hill and Donaldson looked at 374 applicants for 34 staff grade posts in the Northern Region in 1989 and 1990. Five of these were in A&E. These five posts received 43 applicants. Twenty-eight per cent had a full higher qualification with a further 38 per cent having a diploma or the first part of a higher qualification. Considering the applicants for all specialties, 90 per cent qualified outside Europe and only 10 per cent were women. The mean age of appointment was 43 compared to the mean age of consultants on appointment of 37.[21] Another survey of staff grade doctors (in all specialties) and published in 1994 showed that 70 per cent were overseas graduates. Of these, 85 per cent were male but of the UK graduates only 33 per cent were male. Slightly more UK graduates had a higher qualification than overseas graduates (61 per cent vs 53 per cent). The most common reasons for obtaining a staff grade post was the inability to obtain a training post and having no other option if the doctor wanted to stay in hospital medicine or within their chosen specialty but some specifically chose a staff grade post because of their domestic circumstances or to avoid on-call.[22] These surveys give a picture of staff grade doctors as being mostly male overseas graduates who have, for one reason or another, failed to progress in their chosen specialty with a number of female UK graduates who have chosen the post to fit in with their domestic arrangements but 'some women doctors may have entered it because of a local deficiency in flexible training arrangements'.[22]

In 1996, 40 staff grade appointments were made in A&E. The previous positions held by the appointees are shown in Table 8.2.[23]

Table 8.2 Staff grade appointments in A&E 1996

Appointments committees for SG	42
No. cancelled as no suitable candidates	4
No appointment made	5
Appointments made	40
Previous position of appointees	
Locum Registrar	2
Locum SR	1
Locum staff grade	4
Staff grade	13
SHO	7
Medical Officer	5
Clinical assistant	5
GP	2
Registrar	1

Source: Reference 23.

It has been reported that staff grade doctors have problems. One reason is that 'a number of doctors have entered the grade without a clear view of its purpose and the prospects for progress. ... Others may have perceived it as a mechanism to secure additional training in a particular specialty which would make such progress possible by gaining further experience and higher qualifications.'[22] Another problem is that the exact role of staff grade doctors has not been defined. The SCOPME study reported: 'there is confusion among doctors about the status of the staff grade which could contribute to disillusion and dissatisfaction'.[22] and it was reported at a BAEM executive meeting that the 'chief problem that SG doctors were encountering was establishing their position in the hierarchy of departments. This has proved more important to them than remuneration'.[24] In practice staff grade doctors in A&E may be used in two roles. Most probably work as a middle grade doctor[25] and join with any registrars to give middle grade cover to a department where the majority of new patients are seen by an SHO, but in departments with inadequate numbers of SHOs they may be used as an SHO. This may be a cause of dissatisfaction.[26] It has also been said that they should not be reduced to performing uninteresting and repetitive work.[27] The problem is that nobody wants to work as a permanent SHO or to do the uninteresting and repetitive work! It is not easy to get information about the group as most are not members of the BAEM or FAEM and tend not to get involved with medical matters outside their workplace. However one survey (with only a

36 per cent response rate) showed that 60 per cent were happy with their lot.[25]

In 1995 when SHO posts were proving difficult to fill the Department of Health agreed to lift the 10 per cent ceiling on staff grade posts for A&E,[28,29] and since then numbers have increased (see Figure 8.1).

In 2001 the Forum for Associate Specialists and Staff Grades in Emergency Medicine (FASSGEM) was formed and the following year the BAEM formulated a policy on NCCGs emphasising their need for study leave and for personal and professional development. It also recommended that they should be encouraged and given time to participate in teaching, management and the mentorship of SHOs. In addition it recommended that NCCGs should not have to spend more than 40 per cent of their working time in antisocial-hours work.[30] Unfortunately there is a discrepancy between doctors (not just NCCGs) who do not want to work more than 40 per cent of their time in antisocial-hours and patients, 65 per cent of whom arrive in A&E during antisocial-hours. Another genuine complaint of Staff Grade doctors is their remuneration which has lagged a long way behind that of Specialist Registrars (with their payments for additional hours) and general practitioners.

These grades are very important to the specialty and to the smooth running of A&E departments and FASSGEM has representation on both the BAEM Executive Committee and the Faculty Board.

Non-standard grades

When there was a ceiling on staff grade posts, some Trusts got around this by appointing doctors to grades outside the standard grades. These are usually known as trust grade posts and can be considered as staff grades. Since the lifting of the ceiling, the requirement for these non-standard grades with local rather than national terms and conditions of service has passed but appointments to these grades continue. The differing names can cause confusion. Thus the Classified Section of the *BMJ* for 9 December 2000 contained advertisements for Trust Doctors (South Tyneside District Hospital, Harrogate District Hospital), Emergency Medical Officer (Homerton Hospital), Clinical Fellow (Manchester Royal Infirmary) and Emergency Physician responsible for acute admissions responsible to emergency department consultants (Worthing Hospital).[31] In 2001/02 nearly a quarter of non-consultant posts advertised (in all specialties) were for non-standard grades but the number of such posts is not known, as the DoH does not keep statistics.[32]

Trusts advertising such posts expect them to be filled with doctors from outside Europe and doctors coming to the UK for training.[32] However at present A&E is popular and there are many UK-trained doctors who plan a career in the specialty. If they are unable to enter a training programme, either because they do not yet have a higher qualification or because they do not have enough experience to be short-listed, they apply for a post of Clinical Fellow as a Staff Grade appointment appears to them to be a dead end.

While the profession does not like the idea of permanent non-consultant career grades, Trusts and the government do like them as they are cheaper than consultants and are more flexible. A new specialist grade was proposed by the NHS Executive in 2000[33] and was discussed again in the NHS Plan: 'Expansion on this scale also creates the opportunity to ensure that there is a clear career path for all senior doctors. We have examined two options here. The first would involve expanding the number of non-consultant career grade doctors, often on trust specific contracts. This option would allow the NHS to get more fixed clinical sessions from senior doctors without competing with private practice, and it will be kept under review. The second option is to make hospital care a consultant delivered service, where there is a clear career structure so that doctors have certainty about how they will progress and where contractual obligations to the NHS are unambiguous. It is this option that both the professions and the Government support in principle. Its implementation, however, will depend upon a new consultants' contract.'[34]

If past experience is any guide, this new specialist (if introduced) is likely to be thought appropriate for A&E.

9
Junior Staffing of A&E Departments

Providing emergency care in all specialties has always been a problem because the service needs to be provided 24 hours a day, 365 days per year. Consultants have been reluctant to commit themselves to work outside normal working hours except for being on call for those problems which need their special expertise and hospitals have been reluctant to allow consultants' time to be involved in emergency care for fear that the outpatient or inpatient waiting list grows excessively. Thus the care of emergencies in all specialties has largely fallen to junior doctors. This, in itself, has tended to exacerbate the problem as junior doctors working excessive and antisocial hours, believe that by the time they become consultants they have 'earned' the right to more civilised working hours and conditions.

The staffing of casualty departments in the 1950s and 1960s has been described in Chapter 1 and it will be seen that it depended largely on very junior doctors and in many cases on pre-registration house officers (PRHO). Even when full-time casualty SHOs existed, house officers were often expected to cover at night and weekends.

Two facts are generally accepted as true. First A&E is recognised as providing excellent training opportunities and experience *as long as there is adequate supervision and teaching* (the qualifying statement often being forgotten in practice). In 1970 in the letters pages of the *BMJ* it was stated that A&E 'is the most valuable halfway-house between the ward and the world, no matter what a doctor intends for his future practice'[1] and that 'it goes without saying that there is immense value in casualty training. This applies to every practising doctor'.[2] Second, because of the wide variety of A&E work and the fact that patients are seen early in the course of their disease, errors are easily made and experienced doctors are required.

Because much of the work was considered trivial and because of the out-of-hours nature of much of it, it continued to be done by very junior doctors and Clarkson considered, that as long as there was supervision, this was the best arrangement.[3] Most people, however, recognised that it was not ideal. Corbishley, in a letter to the *Lancet* felt that pre-registration house officers were neither experienced enough nor fast enough to work in casualty[4] and the Platt Report said that '... at many hospitals recently qualified staff when serving in the casualty department undertake responsibility which they would not have on the wards or out-patient clinics'.[5] Garden, describing the department in Preston in 1965 said that the futility of staffing the casualty department by a rota of newly qualified resident officers was recognised before the introduction of the NHS, so a fully qualified junior surgeon was appointed to work in the department.[6] In Dundee at around the same time, GPs were employed in the casualty department. One reason for this was to 'give continuity between one lot of casualty officers and the next, helping to avoid the constant repetition of organisational errors'.[7] In other hospitals, senior casualty officers were appointed for the same reasons.

Gradually there was a move to employ more experienced doctors in casualty. Sir Robert Platt felt that 'junior staff in casualty departments... should be fully registered medical practitioners, preferably in their third or subsequent years after qualification. In the larger departments there is a place for senior house officers but this work is of such responsibility that registrars or doctors who have had registrar experience are to be preferred.'[8] Very few departments were staffed solely by registrars but gradually the idea spread that a provisionally registered HO should never be the sole casualty officer on duty at any hospital, by day or night[9] and in 1968 the Department of Health recommended that the standard staffing of an A&E department should be by post-registration doctors: 'There is evidence that a major accident and emergency department staffed by an establishment of five or six doctors in post-registration grades can be practically self-contained.' They further recommended that: 'every effort should be made to avoid employing pre-registration doctors in accident and emergency departments'. House officers were not banned but it was recommended that where they were employed, they should be closely supervised. In particular, hospitals were asked to avoid situations in which junior staff employed in other parts of the hospital, were expected to take responsibility without proper training or supervision.[10]

Even when more senior grades did work in the A&E department, they tended to do so more during normal hours leaving the majority of the night work to be done by the more junior doctors. Table 2.2 shows the average working weeks of different grades in 1969. Following the appointment of the first A&E consultants in 1972, the CSA advised on the staffing of an A&E department: 'For a department dealing with at least 25,000 new patients per annum and providing continuous 24-hour cover, the minimum staff should be one consultant and four or five senior house officers. The ideal staffing for such a department would be two consultants, one senior registrar and five to six senior house officers.'[11] This number of SHOs repeats the recommendations made by the Department of Health in 1968.[10]

House officer posts in A&E slowly decreased but there were still some in the early 1970s with a current A&E consultant having his interest in the specialty started during a three-month A&E post in Newcastle in 1971.[12] In some places PRHOs continued to cover A&E at night. At the Hallamshire Hospital, Sheffield the PRHOs were not all replaced by SHOs until 1983[13] and at the West Cornwall Hospital, Penzance, PRHOs continued to cover one afternoon per week until 1995.[14] Figure 9.1 shows the number of house officers in A&E in England according to the DoH statistics. This shows a sharp decline in numbers between 1973 and 1977. I cannot explain the continuing small number of HO posts in A&E extending into the 1990s except by error or by counting the orthopaedic

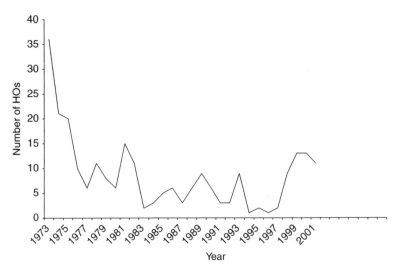

Figure 9.1 Number of HOs in A&E in England 1973–2001
Source: www.doh.gov.uk/stats/history.htm

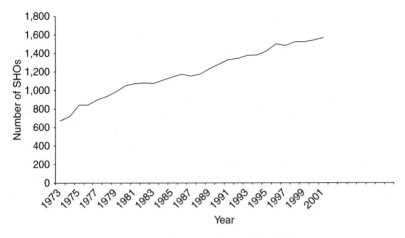

Figure 9.2 Number of SHOs in A&E in England 1973–2001
Source: www.doh.gov.uk/stats/history.htm

PRHOs of those departments where A&E was run by orthopaedic surgeons. Thus the staffing for North Devon District Hospital A&E department was described for 1993 in the 1996 BAEM directory as one staff grade, one clinical assistant and four SHOs. It goes on to say: 'run by 3 consultant orthopaedic surgeons. Same staff for trauma service plus 1 pre-reg house officer'.[15] SHO numbers are shown in Figure 9.2.

Difficulty in obtaining junior staff

The difficulty in obtaining junior doctors has been a constant theme in the history of A&E for the past 50 years. There can be no doubt that the lack of consultant supervision, teaching and career prospects prior to 1972 was a significant factor as was demonstrated by the lack of difficulty in obtaining doctors in the few departments which were well supervised and the improvements seen when the first consultants were appointed. There were other problems with working in A&E. The posts were thought to be of low status and the work was (and is) irregular and involves night work. There was said to be a flavour in A&E of caring only for trivia with all interesting or significant conditions handed over to other services.[16] Accommodation and facilities were often poor. Another problem was said to be the nursing staff. In the absence of consultant involvement, the senior nurse in the department who had a wealth of experience took on a leadership role and supervised the junior doctors. In most places this was probably done with tact and the SHOs were grateful for advice on problems which they had never previously encountered but some relationships were clearly difficult. Problems arise when inexperienced doctor and experienced nurse disagree and it is the less-experienced doctor who has the legal responsibility for making a decision. The attitude of nurses was reported as contributing to the unpopularity of A&E and Bowers in a letter to the *BMJ* said that he had been advised by his Defence Organisation in 1975 that 'it is now generally accepted that casualty officers work under the clinical supervision of nursing staff'.[17] Another person describes the tyranny of senior nursing staff.[16]

By the late 1970s many departments had consultants and the first senior registrars had been appointed thus giving a career pathway similar to other specialties so staffing should have been easier but problems have remained. Some periods have been more difficult than others and different factors have applied at different times. Almost all of these have been outside the control of the specialty (though, given an overall shortage, factors making some departments popular and others unpopular are certainly, in part, within the control of the consultants in those departments).

In 1977, a letter to the *BMJ* stated: 'Repeated recent advertisements for posts in accident and emergency work only have elicited no response or have attracted an occasional overseas graduate with no previous experience of work in this country.'[18] There were problems in 1987/88 when it was reported that, nationally, 18.5 per cent of SHO posts were unfilled[19] and further difficulties in 1994/95 with a 14-per cent vacancy rate in March 1995.[20] As with all shortages of junior doctors, the spread was not even and in the West Midlands region in 1987 the deficit was 37 per cent[19] and in 1995, Hartlepool, in mid-January, was facing the prospect of no SHOs for the beginning of February despite advertising in the UK and Europe[21] and in August had five SHOs, none of whom had previously worked in the UK.[22]

The main problem is that as A&E employs more SHOs than any other specialty, the effects of an overall shortage bite hardest in A&E. In 1993 in England and Wales, A&E employed 1.6 per cent of all consultants, 1.5 per cent of all higher trainees but 11.4 per cent of all SHOs. In 1993, 3675 UK graduates registered. There were 1695 SHO posts in the UK each of which is usually held for six months. A&E therefore would need to recruit 92.2 per cent of all UK graduates to fill the established posts without recruiting foreign graduates.[23] Much of the problem has been an overall shortage of doctors. By about 1980, numbers of overseas doctors, many from the Indian subcontinent, who had supported A&E

departments for many years, started to reduce. The main reason was the 1978 Medical Act which limited the number of foreign qualifications acceptable for full registration with GMC and restricted to five years, the ability to practise with limited registration. In 1985 new arrangements for immigration of overseas doctors were introduced which further restricted numbers. Training also improved in their own countries. Between 1976 and 1986 overseas SHOs reduced from 57 per cent to 34 per cent and overseas registrars from 57 per cent to 45 per cent.[24]

By the late 1980s these doctors started to be replaced with doctors from the European Union but they too, started to reduce in numbers in the late 1990s.

After many years of concern about junior doctors' hours, the New Deal was agreed in 1991.[25] This said that immediate action should be taken to ensure that junior doctors' hours of work did not exceed 83 hours for doctors working on-call rotas, 72 hours for doctors working partial shifts and 60 hours for doctors working full shifts. By the end of December 1994, hours should not exceed 72 hours for doctors working on-call rotas in hard-pressed posts, 64 hours for doctors working partial shifts and 56 hours for doctors working full shifts. The actual hours working should be a maximum of 56.

This did not have a great effect on A&E directly. Almost all A&E SHOs worked a full shift system and probably few worked more than 56 hours though sometimes figures may have been inflated to ensure an enhanced salary and make the posts more attractive. In fact A&E probably gained, as 25 new consultant posts were created.[26] Indirectly however the New Deal contributed to staffing difficulties as many additional SHO posts were established in other acute specialties as the only way to reduce hours. This led to a larger pool of posts with no additional manpower to fill them and so more unfilled posts.

From the mid-1970s finance was a factor as junior doctors had started to be paid overtime. A doctor working a one-in-two rota was on call for 104 hours per week and a doctor working one-in-three was on call for 83 hours. On occasions, doctors in medical and surgical specialties would be up and working all night but many of these hours were spent in bed or relaxing in the Doctor's Mess. A&E SHOs working shifts may have been rostered for only 40 hours per week but all these hours would be spent working. This suited some doctors who were studying for exams but a doctor working in A&E would earn considerably less money than doctors working elsewhere.[18,27] In 1992 junior doctors' pay became related to their work pattern with overtime being worked on full shifts being paid more than overtime being worked on call or with partial shifts.

Postgraduate training for general practice was recommended in the Todd Report in 1968[28] and three-year vocational training with two years in hospital posts and a year in general practice became common in the 1970s and a legal requirement in 1982. Most hospitals set up organised GP training schemes but some doctors construct their own. Most schemes contain A&E posts.

By the 1980s most of the SHOs in A&E were intending a career in general practice but this became a distinct disadvantage when general practice became less popular as a career with fewer potential GPs wanting posts in A&E. Table 9.1 shows the percentage of medical school graduates choosing general practice as a career.[29] The popularity of general practice now seems to be increasing again.

There had been a threat in 1985 to remove the compulsion for candidates for the FRCS to do six months of A&E but initial plans by the Royal College of Surgeons were rescinded.[30,31] However this became a reality in January 1994 when the exam regulations changed. A&E did, however, remain the only specialty to be recommended. By this time only a minority of A&E SHOs were surgical trainees and although this move by the Royal College of Surgeons was unpopular in A&E, its effect on staffing was probably very small.

Most junior doctors choose a post in A&E in their first year after registration and a final contributor to the shortage of SHOs in the mid-1990s was a tendency for junior doctors to spend a year or two in Australia or New Zealand immediately after registration and to obtain A&E experience there. Figures on emigration of doctors (temporary or permanent) are very difficult to obtain but there is a strong impression that this occurred.[32]

Table 9.1 Percentage of medical school graduates choosing general practice as a career

Year	%
1974	33.6
1977	32.9
1980	36.6
1983	44.4
1993	25.8

Source: Reference 29.

By 1997 the problem seemed to be resolving. In February 1997 only 32 out of 1291 (2.5 per cent) of posts were unfilled[33] and in February 1998 the figure was 16 out of 1375 (1.2 per cent).[34]

The effects of staff shortages in A&E are probably worse than in other specialties. Many other specialties can, to a greater or lesser extent, reduce their workload by cancelling clinics or operating lists thereby freeing other doctors to cover the emergency workload but this cannot happen in A&E. The effects of shortages of junior doctors are more than just the cause of a major administrative headache. The *BMJ* in 1979 noted: '... Without an adequate complement of staff mistakes may be made – with particularly disastrous consequences in the accident and emergency department – when doctors are exhausted from excessive working hours or have too many claims on their attention. ... Shortage of staff also works against good communication with patients (and their relatives).'[35] More senior staff have to act down which can hamper the training of registrars,[36] and doctors who might normally find difficulty in obtaining a post are employed as 'better than nothing'. In February 1995 a questionnaire to consultants in charge of departments brought 148 replies. There was a 13 per cent vacancy rate and 33 per cent of the consultants had reservations about one or more appointment.[37] There are financial costs in covering the department with GPs working sessions or with agency locums. Junior staff may also be reluctant to work additional hours in their own departments if they can earn more by working for an agency doing locums in other departments![36,38] While most locums are competent (and usually more experienced than the SHOs they are replacing), there are problems with maintaining a service on short-term locums who do not know the department, hospital or area and it is difficult to ensure the standard of their practice.

Trying to solve the staffing problem

The possibility of using compulsion (in various ways) to staff departments has often been raised. In 1966 the BMA debated a motion that GP trainees should be recommended to do a term of duty in A&E or an orthopaedic department. The orthopaedic surgeons preferred their departments to be staffed by free rather than pressed men and others felt that spurious educational reasons were being recommended to staff short-staffed departments. The motion was lost[39] but this did not stop the argument. Rigby Jones felt that if A&E was a prerequisite for general practice, then not only would this produce more SHOs but also every GP could then do two weeks of A&E per year.[40]

There was a similar discussion in 1971 when the BMA debated that experience in a properly supervised emergency department should be part of the training of every doctor. This too was lost. Not only was there objection to conscription but also it was said that there were only about 30 departments in the country suitable for training.[41] Since then there have continued to be demands from A&E doctors that A&E posts should be made compulsory for all doctors, or that a six-month post in A&E should be a prerequisite for the MRCP exam.[22,42–44]

The value of six months A&E experience for doctors practising in clinical specialties is not in doubt but they need to be encouraged rather than coerced. The Lewin Report said: 'It is surely reasonable to say that all doctors should have a responsibility to the accident and emergency services whether in general practice or in hospital practice, and whatever their specialty' and recommended that trainee physicians as well as GP vocational trainees should be encouraged to spend six months in A&E.[45] The Royal College of Physicians will accept six months of A&E as suitable experience for sitting the MRCP exam.

Difficulties in obtaining SHOs led in 1986 to the introduction of nurse practitioners in A&E at Oldchurch Hospital. This experiment was a success. There was further encouragement to use nurses to relieve junior doctors by 'The New Deal'.[25] The shortage of A&E SHOs in the mid-1990s was more incentive to use nurses in roles traditionally done by doctors and by mid-1996, 36 per cent of departments were using nurse practitioners with more planning to introduce them. There is a wide variation in the length, content and academic level of training of nurse practitioners with much variation in what they do. In the same survey 84 per cent of nurse practitioners ordered X-rays, 36 per cent interpreted X-rays, 68 per cent prescribed over-the-counter drugs and 54 per cent prescribed some prescription only medicines.[46] This is a role which is bound to increase with increasing pressure on A&E departments to take over the management of other problems and government pressure for extending the responsibilities of nurses and paramedical staff.[47] This could significantly alter the role of junior doctors in A&E.

During the 1995/96 shortage of SHOs, concerns were so great that there was a meeting at the Department of Health with Gerald Malone, Minister of Health in the chair[48] and the BAEM also produced guidance.[49] The main practical assistance from the DoH was that the ceiling on the number of staff grade doctors in A&E was lifted. This was useful but would not solve the problem, as there was a shortage of suitable applicants for these posts too. This was also an expensive option as the Postgraduate Dean pays half of the salary of an SHO but the trust has to pay the whole of the (larger) salary of a staff grade doctor. Other suggestions from this meeting and the BAEM advice were:

- Making the department more attractive to work in – improve facilities, accommodation, catering, educational programme, access to study leave, rotational posts;
- trying to decrease the workload – referring primary care patients back to GP (not always easy) and ensuring admissions to specialist units went straight to that unit rather than through A&E;
- ensuring that the work done by a doctor was appropriate by employing for example, phlebotomists to take blood and introducing nurse practitioners;
- better advertising in the UK and abroad.

Who are the A&E SHOs?

I have discussed the origins of A&E SHOs at different times in general terms and this is based on extensive personal experience and from informal conversations over many years with colleagues from the whole of the UK. However the amount of factual information is very limited.

Stewart did two surveys of A&E departments in March and December 1995 and obtained response rates of 73 and 92 per cent respectively.[20] Table 9.2 shows the origins of A&E SHOs. The majority are from the UK. It has always been considered that posts starting in August are easier to fill than those which start in February as many doctors prefer to do A&E as their first post-registration job. These figures confirm a higher proportion of UK doctors in the August to February post. Table 9.3 shows how many SHOs were on rotation. It seems that 32–40 per cent of posts are on rotations and about half of these are rotations for general practice. Table 9.4 shows the career intentions of doctors working in A&E. The low percentage wanting to do general practice confirms the unpopularity of that specialty at that time.

Table 9.2 A&E SHO country of origin in 1995

	March (%)	December (%)
UK	71	86
EEC	15	7
Others	15	6

Source: Reference 20.

Table 9.3 Percentage of A&E SHOs on rotational posts in 1995

	March (%)	December (%)
GP rotations	19	16
Surgical rotations	10	9
Medical rotations	5	3
Other rotations	6	4

Source: Reference 20.

Table 9.4 Career intentions of A&E SHOs in 1995

	March (%)	December (%)
Surgical	17	17
Medical	9	14
GP	20	18
A&E	6	8
Others	4	7
Do not know	7	11
Unknown	36	13

Source: Reference 20.

What does a junior doctor in A&E do?

It would be reasonable to imagine that the manpower of an A&E department depended on the workload and that it would be planned rationally but this is not the case. In 1956 Lowden noted that a few larger non-teaching hospitals had more than one casualty officer, usually a registrar or SHO and an HS but that staffing levels depended more on custom than reason with some hospitals employing two, doing no more work than hospitals with one.[50] Fourteen years later, the Accident Review Committee noted: 'There seems to be little correlation between the number and experience of medical staff in the accident department and the size of population which the department serves'.[51]

In 1999 in a postal survey of A&E departments, Cooke looked at the number of medical staff of major departments per 3500 new patients.[52] The results are shown in Table 9.5 and show much variation.

Over the past 30 years there have been a number of staffing recommendations for A&E departments but these are largely based on

Table 9.5 Staffing of A&E departments 1999

Grade	Staff (WTE) per 3500 new patients	
	Average	Range
Consultants	0.16	0.06–0.78
SpRs	0.09	0–0.88
NCCGs	0.19	0–1.68
SHOs	0.53	0–3.11

Source: Reference 52.

experience. The first recommendation was from the Department of Health: 'There is evidence that a major accident and emergency department staffed by an establishment of five or six doctors in post-registration grades can be practically self-contained.'[10] In 1969 the BOA in a survey found that the mean number of new patients seen by a WTE casualty officer was 6250 per year.[53] As noted in Chapter 2, they recommended that a department needed four career grade doctors to provide a 24-hour service (or five if one accepted a 40-hour week). They therefore recommended that a department seeing 25,000 new patients per year should have five doctors with an extra one for every 6250 new patient attendances.

There are very few studies looking at the work of an SHO. In 1961 a Scottish study found that house officers spent an average of eight minutes to see a new patient that they did not treat but that this increased to 23 minutes if they had to treat the patient (presumably simple treatments such as suturing or incision and drainage of abscesses but this is not stated). Return patients took five minutes. They estimated that an extra 5 per cent would be required for other duties for example telephone calls. They also believed that more experienced doctors would be faster.[54]

In 1971 the Joint Working Party on the Organisation of Medical Work in Hospitals followed some A&E doctors who were on duty for an average of 69 hr 37 mins per week. Of this they worked 29 hr 4 mins (41.8 per cent). The time spent in clerical work was 9.5 per cent. As has been discussed in Chapter 2 they received minimal training.[55]

By 1995 the percentage of the working day spent actually working had increased to 78 per cent of which 18.5 per cent was taken up with non-medical tasks. Tham and colleagues found that the average time for a walking wounded patient was 10.4 minutes; for a child it was 10.6 minutes and for a trolley patient it was 27.3 minutes (but this only included one patient requiring resuscitation). They found that doctors could see

3.3 patients per hour or 4.2 per working hour[56] (though this will obviously depend on the case-mix and how many patients are treated by A&E doctors rather than being referred to another specialist). There is no evidence as to whether this is the correct time to spend with a patient and whether a longer time should be permitted to allow better communication, more patient satisfaction and so on.

Assuming that a doctor sees three patients per hour and has four hours teaching per week, five weeks of annual leave, ten bank holidays and 30 days of study leave per year, then he or she can see a maximum of 6396 patients per year for a 56-hour week or 4428 patients for a 40-hour week. Inevitably there will be much variation in workload and if there are not to be enormous queues on busy days, there need to be some quiet times. These figures fit in very well with the CSA Way Ahead document published in 1992 which recommended one SHO per 5000 new patients for a 60-hour week and the revised recommendation in 1997 of one SHO (or equivalent) per 4000 new patients[57] for a 40-hour week.

10
Primary Care in A&E

As has been seen, before the NHS, many families relied on the casualty department to obtain primary care. Although following the 1911 National Insurance Act, workers could obtain free care from their general practitioner, this was not extended to their families. With the start of the NHS, it was expected that hospitals would cease to provide primary care, as all patients would be able to obtain free treatment from their general practitioner. Clarkson reported that at Guy's Hospital there was a 20 per cent reduction in the numbers of sick attending and a 60 per cent fall in the number of children[1] following the introduction of the NHS. However the NHS system of capitation payments offered the GP no incentives for discouraging patients from attending A&E and a GP has written: 'Instead GPs became increasingly concerned that any shift of demand away from A&E would lead to increased GP workload without parallel increases in resources.'[2]

Patients continued to come in large numbers and in a well-known paper, Fry analysed the work of a casualty officer at Kings College Hospital in 1958 and concluded that: 'The present day casualty department has no specific function and runs as a "general practice". It is misused to a considerable extent by the public and by their doctors.'[3]

Patients with primary care problems have been called many things over the years including casuals, inappropriate attenders and primary care patients. A major difficulty has been in trying to define the group and this difficulty results in the proportion of primary care attenders being variously estimated as being anything from 6.7 per cent to 64–87 per cent.[4] Three possible definitions[5] are:

1. Somebody who could be effectively treated in general practice. For example, not likely to require any investigations and not likely to require hospital admission.

2. Patients who are not urgent and who do not require the attention of an A&E doctor. Some people have tried to define by length of presenting symptom e.g. a patient whose symptoms have persisted for more than 24 hours without worsening and could wait for 24 hours before being treated.
3. Somebody who presents with symptoms that are on a pre-classified symptom-based protocol.

Even where authors have tried to define the group, the accuracy of the assessment must be open to question, for example, if they have relied on an SHO with no experience of general practice to decide whether a patient is suitable to be treated by a GP. There are further problems: many rural GPs may suture wounds whereas their urban colleagues do not. Is a wound requiring suturing, a primary care problem? Definitions of primary care problems will change with changes in medical practice: in 1958 Crombie, a GP in Birmingham, felt that 80 per cent of casualty attenders could be dealt with in GP assuming that nitrous oxide anaesthesia could be given in general practice and that GPs had facilities for minor surgery and suturing.[6] In reality, there is a significant overlap between A&E work and primary care. A patient with a neck injury sustained in a road accident is entirely appropriate to be treated in A&E but many such patients are also seen in primary care.

Some authors have implied that patients with minor injuries have primary care problems but any classification based on diagnosis is not appropriate. A person falling 15 metres may escape with bruising only, but no one would dispute their need to be seen and assessed in A&E.

There have been numerous studies,[7-11] good and bad, over many years, looking at why patients attend A&E with primary care problems but they all suffer from the fundamental problem that they depend on asking the patient (or carer), who is likely to phrase the reply depending on the setting in which the question is asked. Thus patients seen in hospital are more likely to say that they could not get hold of their GP or were unable to make an appointment within a reasonable time whereas patients asked the same question by their GP are more likely to say, 'I didn't want to disturb you, doctor.'

However the reasons will usually boil down to one of the following:[2]

1. The patient's situation when the need arises;
2. The perceived availability and accessibility of GP services;

3. The patient's or advisor's view of the urgency and the type of treatment which is likely to be needed;
4. The perception of the costs and benefits involved in attending A&E or GP.

The situational reasons include commuters, visitors and holidaymakers taken ill away from home and thus their GP; students and others who have moved to a new area and not yet registered with a GP. Many patients do not realise that they can register with a GP as a temporary patient.

The availability of a GP will vary with the time of day and day of the week. Rigidly applied appointment systems may make the GP less available to the unassertive patient. The hospital may be nearer than the GP surgery and thus more readily accessible.

A common response to questions as to why a patient attended A&E rather than see their general practitioner is that they feel that they need an X-ray or that the GP would have sent them to hospital anyway. Some patients want a second opinion or feel that they are already under the care of the hospital. The general practitioner is often blamed for not providing a service for emergencies but some of the 'inappropriate' work of an A&E department is caused by inadequate hospital services for patients with semi-urgent problems. Blackwell in 1962[7] described two patients at risk of suicide who had been asked to wait eight weeks for a psychiatric outpatient appointment and a child with hip pain who would have had to wait eight weeks to see an orthopaedic surgeon if he had not come to A&E. Such problems could just as easily have happened 30 years later.

Among the costs can be included that a patient who dials 999 gets free transport to hospital and may be able to persuade the hospital to arrange transport home whereas transport to the GP is the patient's responsibility.[12]

By 1960, the NHS had been in existence for 12 years and it was felt inappropriate that casualty departments should continue to see patients with minor problems. An article on hospital planning said that it was questionable whether hospitals should continue to treat casuals[13] and the Nuffield Report[14] looked at departments' ways of dealing with 'inappropriate attenders' as one of its marks of quality. A department scored ten for a 'polite notice, backed up by informing GPs of function of the department and asking their cooperation, as well as "friendly" persuasion by the staff' but scored zero if they made no attempt at regulation and yet still complained about the department being overloaded with trivia.

The comment: 'the whole lay-out and organisation of the department seems to be geared to turning away the "casual attenders" ...' about a hospital was a mark of approval. It said that 'each hospital at present tends to have its own conception of the arrangements it can make (rather than of what is necessary) for the reception of casual patients ...' and warned of casualty departments: 'if they are organised to ... absorb all demands on them, including that of being an alternative to general practitioner consultations, they will certainly not lack for customers'. For a solution, it suggested that 'there is a need for the fullest consultation between the hospitals and local medical committees (and other appropriate bodies) as to how general practitioners can help to relieve the hospital of the burden of such cases' (relatively minor, non-urgent conditions). It recommended that the question of employing general practitioners as clinical assistants should be explored not only because of the shortage of junior staff 'but because such a move would strengthen the essential links between the hospital and community services'.

The Platt Report[15] also believed that too many minor problems were hindering the treatments of accident victims who were the appropriate users of the A&E department: 'It was widely agreed that one reason for the inadequate treatment of injuries was the crowding of many departments with patients who could have been treated by general practitioners.' The Royal College of General Practitioners said 'Three-quarters of the casualties who for one reason or another go *direct* to hospital could well be cared for by their family doctors – they are really general-practitioner patients in the wrong place and they should be encouraged to go to their family doctor wherever possible.'[15]

A similar view that patients with minor problems were interfering with the care of 'real' casualties were expressed in the correspondence columns of medical journals. Garden was of the opinion that the casualty department should provide privacy and protection for those who have suffered sudden injury or illness and no others should be allowed to cross its threshold.[16] Murley felt that the problem was the NHS itself and that 100 per cent free-at-the-time medicine with failure to differentiate between the urgent and vital and the trivial was the chief reason why the NHS was cracking up.[17]

As a solution, Platt recommended that: '.... an experienced general practitioner "on the door" of a "casualty" department for the purposes of "screening" cases might have excellent results' though 'the primary task of general practitioners, so far as this service is concerned, is to deal with minor accidents and emergencies at any time of the day or night'.

The report was, however, realistic and said that '... some separate provision must be made, at least for the time being, to deal with patients suffering from minor non-traumatic conditions'. It saw a future when, with group practices, there would always be a GP available to see them and the number of such patients will reduce but recognised '.... there will always be some patients of this kind at some hospitals, despite all the efforts of general practitioners and hospital staff to discourage them'. For example renal colic coming on at work or epistaxis when the GP was doing his rounds are emergencies in the patient's eyes and no reclassification by doctors or administrators will prevent such cases presenting themselves at the hospital.[18]

In the late 1960s it was still government policy that minor problems should not be dealt with in hospital. The DoH said: 'Casual attenders not in need of hospital attention are the responsibility of the general practitioner service'[19] and the Todd Report recommended: 'The hospital will not provide primary medical care, except in emergency.'[20]

Another solution was Platt's recommendation that the name '"Casualty Service" should be altered to "Accident and Emergency Service"' presumably to reduce the 'casual attenders'. Some orthopaedic surgeons appeared to believe that 'with the renaming of casualty departments as accident and emergency departments, the amount of work other than true accident work will virtually cease, especially in the provinces'.[21] The *Lancet* was more realistic when it warned that 'a change of name may not do much to abate this problem' of casual attenders.[22]

There are problems when one tries to exclude patients with minor problems. 'From the patient's point of view a condition is only minor when it has been so diagnosed.'[23] One cannot exclude those who walk in as Platt found that self-attenders did not necessarily have trivial problems as 2 per cent were admitted and 61 per cent had further appointments made for them in either A&E or outpatients. Those who attend for a second opinion would appear to be abusing the system but a study of 180 such patients published in 1989 showed that 47 were admitted and two died[24] and another study showed that patients who attend A&E after seeing their GP are as likely to be admitted as those who are referred by their GP.[25]

If patients are turned away, some mistakes are inevitable. In 1967 the case was reported of a casualty officer who refused to see a vomiting patient. The patient later died of arsenic poisoning and the doctor was found negligent. The judge agreed that 'it was not the duty of a casualty officer to see every caller at his department' but said that every case had to be assessed.[26] Even when no medical mishap arose, people turned away from casualty frequently complained which resulted in a reproof to the casualty officer on grounds that the customer is always right.[27] Other disadvantages of turning people away were losing teaching material and loss of good will towards the hospital.[28] The only conclusion could be that an open-door policy is not right but it is not possible to have a closed-door policy and some compromise is necessary.[28] The only person who could decide was the casualty officer.[18]

In 1967 Jennings, an SCO was obviously making a political case for A&E consultants but was probably correct when he said that the problem with inappropriate attenders would be difficult to solve until A&E was recognised as a specialty and got the support of the hospital management committee and its own staff.[29] This was also recognised by the Department of Health who were convinced that when there was good consultant supervision, the number of casual attenders would be fewer.[19] The same circular stated that 'the responsibility for "sorting" patients who present themselves at a hospital into those who need hospital care and those who do not, cannot properly be carried out by other than a registered medical practitioner. It should not be placed on the nursing service'.

Under this system, individuals who attend inappropriately should ideally be seen and advised where and how to seek help. However it may be much less trouble merely to hand out some symptomatic treatment and doctors find it very uncomfortable to say to a patient: 'I know what will help you but you need to see your GP to get it.' It was commonly assumed that 'unless A&E doctors accept the burden of redirecting patients to appropriate services they are cooperating in the misuse of the department and must expect that misuse to increase'.[30]

By the 1970s, things had got no better. Peter London, describing the Platt recommendation on changing the name of the departments, said: 'it would have taken no more than a modicum of cynicism to predict that one consequence of these recommendations would be the purchase and erection of a large number of new signs and that another would be the complete ignorance of the part of the public of the worthy objectives of the committee'.[31] A study on workload said that casualty officers were still 'swamped by large numbers of cases not in need of hospital treatment, such as insect bites, "throbbing" warts, coughs, colds, minor abrasions and headaches'[32] and general practitioners still got the blame. It was said that the workload was increased by 'the growing popularity of appointment systems among General Practitioners'; patient 'difficulty ... in contacting and obtaining the services of their general practitioner at night and weekend' and the limitation of services by some GPs. 'Some general practitioners would undertake the suture of wounds, others, it was said, would not.'[32] An A&E consultant also blamed 'the withdrawal of most general practitioners from the treatment of trauma in any form, even the simple, dressing of wounds, largely because the "nurse" has been replaced by the "receptionist" '.[33]

Not only was there no incentive for GPs to treat patients to treat injured patients but also there was a financial disincentive as they had to provide their own dressings, gloves, syringes and instruments and it had been argued that these should be supplied for them.[34] Other recommendations at the time for reducing the A&E workload included patients receiving transport to health centres and GPs having access to X-rays, knowing that they could have a report within 48 hours.[35]

Problems have always been worse in the inner cities, particularly London. In the early 1990s Inner London averaged 405 A&E attendances per 1000 population compared to the England average of 235 per 1000 population.[36] A contributory factor is that: 'For a number of groups – the homeless, drug abusers, some members of minority ethnic groups, commuters, visitors – GP services are for various reasons, not available. For some groups they may not be acceptable either, and even if they were, it is likely that the street homeless, for example, would be unacceptable to other users of GP services.'[36]

The standard of general practice in some inner city areas is (or was) not good with many single handed GPs, trained in different cultures with poor premises, fewer employed practice staff, no rota systems and larger than average list sizes. Such GPs become demoralised. Some inner city GPs devote much time to private work and may be unavailable to their patients for a good part of the day and inevitably patients of these doctors find their way to the A&E department.[36–38] However despite the observations of the A&E staff, they only saw one side and there was no evidence for a significant shift in workload from primary care to A&E. In Glasgow from 1973 to 1977 A&E workload increased 3.9 per cent per year but GP workload increased 3.4 per cent per year with out-of-hours home visits increasing 11.7 per cent per year.[39] It was pointed out that six times as many patients were seen in primary care as in A&E so that a 10 per cent shift of work away from GP would increase A&E workload by 60 per cent. What was happening was an increase in workload for both primary care and A&E.

It is interesting to know what GPs wanted. A survey of 53 GPs in Birmingham in about 1971 revealed that they wanted A&E to provide:[35]

1. X-rays
2. Reception of trauma of all kinds
3. Facilities for stomach wash outs

4. Suturing of lacerations
5. Excision of cysts
6. Incision of septic cysts
7. Treatment of epistaxis
8. Treatment of foreign bodies in the eye
9. Ability to refer patients to physiotherapy
10. Sterile packs.

An editorial in the *BMJ* in 1974 noted that the out-of-hours GP emergency service, whether provided by GP or deputising service was geared towards domiciliary emergencies and recommended that when GPs form a group practice, they should be required to provide a 24-hour service for minor injuries.[40] Needless to say, this was not well received by GPs. One GP who described himself 'as someone who has frequently had his Sunday lunch ruined putting on a small plaster dressing to a scratched finger ...' wrote: 'To suggest that a general group practice centre be turned into a 24-hour minor casualty department ... is complete nonsense ... .' He also was 'not at all sure that a 24-hour minor casualty service is necessary or desirable ... [as] to encourage an over-pampered public further open access to medical care will destroy even what is left of the little encouragement allowed to people to use their common sense'.[41] More reasoned thought agreed that minor trauma was primary care but felt that emergency minor trauma can be more economically dealt with by centralising it in hospital.[42]

The *BMJ* editorial had taken a rather dismissive view of the minor problems that attend A&E and were reminded by John Scott, the founder of the Oxford Accident Service whose views were usually that A&E departments were for trauma only, that: 'A "minor" emergency is one that happens to other people' and also said that the *BMJ*'s 'reference to casualty attenders as "the rag-tag-and-bobtail of the medical caseload" is a clear indication that attitudes which have plagued attempts to find solutions to this complicated problem have not changed'. He stressed that A&E departments needed more resources.[43] Hardy, an A&E consultant in Hereford, felt that 'there are bound to be some patients whose troubles seem to them to be major and urgent however minor and intrusive they seem to us'[33] and the Court Report, on services for children, stated that in some areas, especially in the inner cities, 'it is considered inevitable that for the foreseeable future, the A&E department will continue to be used as a source of primary care ...' and said that that would need to be taken into consideration in the staffing arrangements.[44]

The problem of increasing numbers of patients requiring advice and treatment for minor problems was not going to go away. Education has never been shown to reduce the number of patients attending A&E though it may be successful in increasing utilisation of alternative services.[5] There was obviously a balance between GPs being resident on call at health centres to provide care for emergencies and being able to completely abrogate responsibility for out-of-hours work.[45] The problem was a joint one: 'Better facilities, badly needed, for the treatment of increasing numbers of patients attending casualty departments must come jointly from general practice and hospitals.'[40] It was clear that GPs and A&E had to work together and the Court Report in 1976 had also made a plea for more rational working relationships between A&E and primary care (and outpatient departments).[44]

The first authoritative advice that special arrangements should be made in A&E for patients with minor problems came in 1979 from the Royal Commission on the Health Service: 'In large cities the local hospital has sometimes been used as a walk-in GP surgery by patients who find it convenient to receive their treatment there. ... Where the tradition of using the department in this way is strong it may be preferable for the hospital to accept its role and make specific arrangements for fulfilling it rather than try to resist established local preferences.'[46] The Mills Report was less certain: 'Any tendency to develop an additional primary care service based on the hospital accident and emergency department should be undertaken only with great caution and only on an experimental basis perhaps in relation to very mobile populations, for instance, in parts of London.'[47] The CSA also felt that primary care in A&E departments should be kept to the unavoidable minimum.[48] There was (and still is) also a feeling among A&E consultants that London's problems are special and should not unduly influence the provision of A&E services in the rest of the country.

The next mechanism to cope with the increasing numbers of minor problems was triage. Triage means sorting and had long been used by the military to sort casualties so that the more seriously injured were treated first. Triage had also been used in A&E departments primarily to sort the injured from the ill or medical from surgical patients when the two groups were treated by different individuals. It had also been used informally to try to avoid the seriously ill and those in pain waiting too long for treatment. This informal triage frequently relied on the receptionist calling a nurse to see any patient who appeared unwell. It has been noted above that in 1968 nurses had been forbidden by the DoH from sending patients away without being seen by a doctor. This was probably honoured in the breach as I am sure that the night sister in my own department was not the only nurse in the country who said such things as: 'If you think I'm going to get *my* doctor out of bed for *that*, you are very much mistaken!' Formal triage started to be introduced in the early 1980s.[49] It involves all patients being seen by a nurse and allocated a priority so that those triaged category one were seen before those in category two who were seen before those in category three and so on. Systems varied but those with minor problems triaged with a low priority could either be advised to see their general practitioner[50] or told that they would be seen but there would be a long wait, as they would only be seen after everybody else. An extension to the idea of triaging patients on arrival at A&E was to advise them to telephone the department to ask whether they needed to be seen. This was pioneered in Preston where the triage nurse might suggest a later visit, a review clinic appointment or that the patient be seen immediately. Casual attenders could be told if a visit would be unnecessary or inappropriate and the triage nurse could make a GP appointment if necessary.[51,52] GPs, receptionists and other community health carers were also asked to phone the triage desk rather than sending patients in. Triage later gained great importance in the NHS such that all departments were forced to introduce it (even if they had no queues) and for a while the percentage of patients triaged within 15 minutes of arrival in A&E was the only way that A&E performance was measured.

The Royal Commission's recommendation that 'it may be preferable' in some cases 'for the hospital to accept its role and make specific arrangements for fulfilling it rather than try to resist established local preferences'[46] was taken up at Kings College Hospital by Jeremy Dale in 1989 in a collaboration between A&E and the Medical School Department of General Practice Studies. Patients triaged as having a primary care problem were randomised to receive care from a GP employed in the A&E department or from an SHO or registrar. GPs requested fewer X-rays or pathology investigations, wrote fewer prescriptions and made fewer referrals to inpatient teams, outpatient departments or for A&E follow up than did SHOs or registrars.[53] They were also more cost-effective.[54] This work was criticised in that vocationally trained GPs were compared with trainees and produced no data to compare GPs with consultants. Consultants should have been better than trainees at dealing with these problems and consultants have the advantage of being able to treat major as well as minor injuries.[55]

Following Dale's early work, the Tomlinson Report into health services in London in 1992 recommended that 'where there is a high proportion of primary care attenders, adapt hospital A&E services in a cost-effective way, for example by including GPs and nurse practitioners amongst the A&E staff so that patients get the service which is appropriate to their requirements.'[36] In February 1993 ten GPs started to work in the A&E department at St Mary's Hospital and they were found to have very similar advantages to those at Kings College Hospital.[56] Similar systems were started in a number of hospitals in London. A postal survey of 33 A&E departments in the North Thames Region produced 31 replies, 16 of which reported a primary care project of some sort.[57] Half of these were funded by Tomlinson money and five were close to the Kings model. This study reported various problems with this model which had not come out of Dale's publications. Some GPs found that this work conflicted with their practice commitments and it was also financially unrewarding as they were paid at clinical assistant rate only and were paid just for their working hours and not for any educational time they put in for A&E. When working as employees of the hospital, GPs lack their usual autonomy. This would be appropriate if they are doing a specialist clinic under the supervision of a consultant but is inappropriate when they are performing their own specialty of primary care. Most primary care work in A&E occurs during antisocial hours and there were problems recruiting GPs for these times. There was some ambiguity of role. Many GPs were looking for a change from their normal practice when doing shifts in A&E and wanted to see some sick and interesting patients rather than the primary care patients. Some GPs thought that if they were just going to provide primary care, this was best done in the patient's own practice with continuity of care.

It has been seen from some of the quotations above that there has been the fear that providing a good service to primary care patients in A&E departments would increase demand[14,30] and the projects of placing GPs in A&E departments needed to decide whether they aimed to provide a better service with the risk of creating more demand with little prospect of more funding or of providing a service aimed at diverting patients elsewhere.[57] In fact, fears that providing more appropriate care would open the floodgates of demand does not seem to have happened.[2,56]

Ten hospitals in North Thames had introduced nurse practitioners. These are cheaper to employ per hour than GPs but as far as the primary care load is concerned, they are slower and see a smaller range of conditions with a caseload containing more trauma. As a consequence they may be more expensive than general practitioners.[2]

An alternative approach in a few hospitals was to appoint consultants to the A&E department with the role of improving services to primary care patients. Judith Fisher was appointed a consultant in A&E with an interest in primary care at the Royal London Hospital in 1993[58] and Iain Robertson-Steele as a consultant in primary care within the A&E department in Wolverhampton in 1994.[59] The role of such consultants was never formally evaluated and in at least one hospital there were problems and confusion over the role of the consultant in primary care in relation to rest of the work of the department.

In 1995, in response to increasing dissatisfaction by GPs with their out-of-hours commitments, the Department of Health established a development fund for GP out-of-hours services and reformed their payment and fees structure. GPs' terms and conditions of service were altered to stress that GPs were the sole arbiters, on grounds of clinical need, of whether and where a patient should be seen. This encouraged GP cooperatives to offer centre-based care as an alternative to home visiting. As a result of these initiatives, the number of cooperatives and primary care emergency centres mushroomed almost overnight. Some primary care emergency centres were located in or adjacent to an A&E department or at least within a DGH. Hallam and Henthorne evaluated GP cooperatives and primary care centres.[60] They studied seven cooperatives running 11 emergency care centres. Two of these were in or adjacent to A&E departments, two were in other hospital departments and five were in community hospitals, four of which had a casualty department or Minor injuries unit (MIU). Most chose not to align themselves to an A&E department.

Advantages to an emergency centre adjacent to A&E were:[2,60]

- Good parking, patients know where it is, well signposted.
- Use of nursing staff, 24-hour security and portering. This can lead to cost savings.
- In a stand-alone centre, a receptionist may be left on her own at night and feel very isolated if the GP goes out on a visit.
- Access to the hospital clinical facilities and equipment. This leads to 'One-stop' care with investigations, second opinions, admissions all on site and a more seamless service.
- Development of greater understanding and working relationships with A&E staff, cross referrals where appropriate.
- Patients may prefer to attend primary care centres attached to A&E. In one study of 20 cooperatives, two primary care centres were attached to A&E or MIU. Between 50 and 66 per cent of callers attended. Eight were based in cottage hospitals where the attendance rate was 4 per cent to 45 per cent. Ten cooperatives were based in other places with an attendance rate of 7–37 per cent.

There is also the advantage of more integration of services. In Croydon the GP Cooperative contracted to see 10,000 primary care attenders with financial advantage to the cooperative and, for the A&E department, the ability to concentrate on the more seriously sick and injured.[61]

Potential disadvantages of an emergency centre adjacent to A&E to the GP cooperative are:

- Fears of increasing workload with work offloaded from A&E and also a fear that the high profile of a hospital site might encourage demand.
- Confusion of roles in minds of patients (24 per cent of patients attending a primary care centre established in a MIU claimed they had attended A&E).
- Risk of over-medicalising minor problems by asking patients to come to hospital.

Sharing a space used by someone else in the daytime (not just hospitals) caused its own problems in that every evening they had to set up their equipment and every morning pack it all away again. Computers do not respond well to this treatment. GPs also found that domestic facilities were poor and there were security problems for their equipment when the primary care centre was not working.

There are also potential disadvantages to A&E of having an emergency centre adjacent to it:

- If a centre is present part of the time, it will attract primary care patients to A&E. Then when the centre was not running, they would have to be seen by A&E staff.
- Some small A&E departments were afraid that if numbers were reduced, they would not be viable.
- When a patient walks in the door, one cannot tell if patient is a casualty or a GP cooperative patient.
- One centre had to be moved as A&E had failed to anticipate the number of patients who would attend the emergency centre especially at weekends and bank holidays and waiting room and nursing staff could not cope.

Centres on the DGH site but away from A&E have the same advantages and disadvantages except that there is less risk of confusion.

About 4 per cent of callers to the cooperatives were advised to attend A&E but it is very difficult to know what effect GP cooperatives have had on A&E departments. In one area the opening of the GP centre led to a fall in A&E workload of 200 patients per month but in most areas A&E attendances are rising but it is difficult to know how much is national trend and how much is due to changes in the primary care emergency system.

Factors tending to increase the A&E workload are:

- An increased propensity of GPs to refer patients they do not know;
- Increased referrals when the cooperatives are overloaded;
- Rural GPs have often done minor trauma work. This might decrease if patients are cared for by a cooperative;
- Difficulties in accessing the cooperative by telephone;
- In some cases A&E may be nearer than a distant primary care centre;
- Some cooperatives encouraging specific referrals, for example, catheters.

But there are also factors which might decrease the workload:

- Agreed policies on referral of primary care patients from A&E to cooperatives,
- More accessible, higher profile GP service out of hours,
- Long waiting times in A&E for low priority patients.

Primary care initiatives within an accident and emergency department were largely confined to London and GP primary care centres within A&E departments were not widespread with the majority of GP cooperatives siting them elsewhere. These were not therefore national plans to deal with primary care problems in A&E. National plans began to emerge in 1997 with the report 'Developing Emergency Services in the Community'.[62] Its starting point was that '... when a person thinks a situation is an emergency, then from the point of view of service response, it is an emergency until there is sufficient evidence to change that view. An emergency is only minor in retrospect.' It also called for better emergency mental health services, improved availability of emergency dental care and complementary services to provide immediate care including 24 hour a day emergency nursing services, community and outreach services to care for people at home.[63]

It also called for an NHS telephone helpline which has developed into NHS Direct. For many years patients, uncertain of the significance of their symptoms, had telephoned their GP or A&E department to ask whether they needed to be seen. (The medico-legal implications of this had worried some within A&E.) In the early 1990s in Preston, as has been seen, this was more formalised. At the same time GPs and their cooperatives were instituting telephone triage as a means of controlling their out of hours work. NHS Direct was to be a national telephone advice line whose aim 'is to provide people at home with easier and faster access to information and advice about health, illness and services so that they are better able to care for themselves and their families. The service provides clinical advice to support self-care and appropriate self-referral to NHS services ...'[64] One of the hopes was that by offering appropriate advice, A&E attendances for minor problems could be reduced. This was introduced as a political decision, without evidence of effectiveness and the fear was that it could increase the workload of A&E and primary care. Symptoms are very common and diary studies suggest that as few as one in 40 of these ever reaches a consultation. If these patients were encouraged to seek help, a percentage of them might be directed towards A&E departments.[65] The effect of NHS Direct on A&E workload is difficult to prove in view of the increasing numbers using the service each year but it has been found that 'changes in trends in use of accident and emergency departments ... after introduction of NHS Direct were small and non-significant.'[66] Other government initiatives which aim to reduce A&E workload, are walk-in centres and additional training for paramedics to enable them to leave patients at home or to refer to other agencies rather than bring all patients to hospital. It is too early to assess the impact of any of these measures.

Up until the 1990s the only systems available for obtaining emergency help were the GP, the A&E department and, in rural areas, the minor injuries unit (MIU) in a cottage hospital where care was still the ultimate responsibility of the GP. With the increasing reluctance of GPs to provide an out-of-hours service,[67] GPs generally opted out of responsibility for care in MIUs many of which became nurse-run units. MIUs were also being opened in cities, usually replacing an A&E department when it merged with a neighbouring department. Patients now have a choice of A&E, MIU, GP services or walk-in centre. '... There is potential for confusion and duplication of services, leading to inconsistent responses, variable quality of care and inefficiency.'[68] GPs also believe that 'it is ironic, for example, that most cooperative-led primary care centres restrict access to people who have been previously triaged by telephone, yet patients with identical minor illnesses can walk into A&E departments (often next door to the primary care centre) to see a doctor with no primary care training'.[68]

The latest thinking on the primary care attender is that 'there is no such thing as an "inappropriate attender"; anyone who has, or thinks they have, some sort of urgent physical, psychological or social need is appropriate. It is the services, the way they are assessed, and the way in which they are delivered that is inappropriate. ... Forget about "managing demand": think instead about responding to need.'[69] There is an acceptance that A&E is not the best agency for responding to many of these needs and there needs to be much greater coordination between different providers. There is a need for collaboration and integration of services between the hospital and outside agencies and within the hospital so that every patient gets the help that they need.[70] In 2001 the Department of Health, in a document 'Reforming Emergency Care', recommended (among other things) an integration of emergency services so that patients would be assessed in the same way using a computerised clinical assessment system (CAS) whether they phone NHS Direct or attend a primary care centre, walk-in centre or A&E department.[71] Patients would then be streamed so that every patient would be seen by the 'right' service for their needs, regardless of where they presented. Some could be discharged after NHS Direct-style advice with others being referred to primary care centres and others would bypass A&E to be seen by specialist teams. In A&E departments patients with minor injuries would be separated from those with major problems and seen in a separately staffed minor injuries unit (MIU) so that their treatment was not delayed by the arrival of a more seriously ill patient. There was an ambitious target for all patients to be seen, admitted or discharged within four hours of arrival and additional resources were to be put into emergency services.

If that were to happen, it would radically alter the A&E department because if the minor workload is removed by MIUs, walk-in centres and primary care emergency centres then A&E would be left with a marked increase in the complexity of its work.[72] This would have implications on staffing and so on. In addition many doctors enter A&E as they enjoy the variety of the work and do not want to have a large section of their work removed.

In fact there have been changes to the plans which will be briefly described in the next chapter.

11
Politics and the Future

Reading this book will show how political decisions have influenced hospitals and their A&E departments ever since the formation of the NHS. However most of the changes occurred with the consent of the medical profession and often as a result of reports written by members of the profession. For doctors and nurses working in A&E departments, these changes did not have much influence on the day-to-day working of departments.

The first political decision that really affected those working at the sharp end was the *Patient's Charter* in 1991.[1] One of the points in the charter was the assurance to patients attending A&E departments: 'You will be seen immediately and your need for treatment assessed.' This implies triage. As has been seen in Chapter 10, many A&E departments had introduced triage in the 1980s as a means of ensuring that the more seriously ill patients were seen early and that nobody died or deteriorated while sitting in the waiting room. The Patients Charter had a big influence on the way A&E worked. Departments where patients never waited long to be seen by a doctor were forced to introduce triage even though they felt it was unnecessary. Other departments that had been using triage for years had to alter their way of working to ensure that patients were seen immediately and there was discussion on whether waiting two minutes was immediate or not. (Immediate was later clarified to mean within fifteen minutes of arrival.) Health authorities started to monitor this which led to systems being changed so that patients were seen by the triage nurse before booking in as this saved five minutes. My own department was forced to install a supermarket-style ticket machine for patients to take a time-stamped piece of paper so that we could time a patient's arrival, exactly. Other departments put a nurse in reception (familiarly known as a 'hello' nurse) so that it could be claimed that every patient was seen by a trained nurse as soon as they arrived.

The Labour Party victory in the 1997 election had a significant effect on the management of emergencies in hospital initially by making things more difficult. The emphasis put by the new government on waiting lists for surgery, outpatients and cancer treatment meant that all available resources and managerial effort went towards elective care, particularly surgery, and emergency care missed out.

This changed in the new millennium and the pointer was *Reforming Emergency Care*[2] published in October 2001. Initially this seemed to be a threat. In the interests of improving patient flow, it recommended triage, with patients being referred directly to the department best capable of looking after them: minor injuries to a nurse-run minor injuries unit, primary care problems to primary care, chest pains to the coronary care unit, fractures to the orthopaedic surgeons, strokes to a stroke unit, and so on. A&E departments would be left with resuscitation. This is very similar to the idea put forward in 1972 (described in Chapter 2) of being able to abolish casualty by each specialty having its own emergency room. The lessons of history are that this does not work and that it is difficult enough to have one department at constant readiness to receive patients, let alone several departments equipped and staffed to cope with all comers. While direct referral, in theory, should be more efficient by removing steps from a patient's journey, there is no evidence that it would be more effective. As noted in Chapter 6, rapid thrombolysis for myocardial infarction can best be done in an A&E department. It is likely that A&E would be best able to deliver other time-sensitive treatments such as those proposed for stroke. A&E departments are usually designed for easy access to X-rays and other imaging modalities whereas wards may not have such good access.

Reforming Emergency Care also stated that by April 2003, 90 per cent of patients would be seen, treated and discharged from A&E within four hours and that by April 2004, this would rise to 100 per cent. This appeared to be an impossible target.

Within 12 months of the publication of *Reforming Emergency Care* many of the main ideas (though not the four-hour target) had been abandoned. Triage was considered wasteful by putting an additional, unnecessary step into the patient's journey and slowing it down. The computerised clinical assessment system was slow and seems to have been abandoned. Streaming of patients into three or four small queues rather than a single queue is inefficient. The four-hour target concentrated managerial minds and suddenly resources became available for A&E departments and so the potential threat turned out to be an advantage. A major problem had been patients waiting many hours (sometimes more than a day) in A&E for a bed to become available on a ward. This had an enormous knock-on effect on the department with staff and space being used to look after patients who should be on a ward and thus not available for new patients. When managerial jobs were at risk if the targets failed to be met, it was inevitable that there would be some innovative solutions such as renaming A&E cubicles as a ward, but overall the effect of the four-hour target has been beneficial to A&E. The four-hour target has not yet been achieved but there has been enormous progress.

In order to help A&E departments to meet the targets, the NHS Modernisation Agency set up a series of collaboratives to spread good ideas. One recommendation was the idea of 'see and treat' or 'greet and treat' in which the majority of patients with minor problems were seen by an experienced emergency nurse practitioner or experienced doctor who could see patients quickly, make decisions and either treat them immediately, or advise them where they can seek help more appropriately[3] (familiarly known as 'greet and street'). This moved the experienced A&E doctor from the major side of the A&E department to the minor side. It undoubtedly speeds up patient through but is inappropriate if a consultant or experienced middle grade doctor is 'queue-busting' on the minor side of the department while decisions on more severely ill patients are being made by SHOs.

There is an intention that A&E workload can be reduced by a significant degree by providing alternatives such as a better primary care service, minor injuries units and walk-in centres. Many 999 calls to the ambulance service are 'inappropriate' and there are currently plans for extended training of paramedics (and others) to become Emergency Care Practitioners who would be able to treat patients in their homes or refer to more appropriate services. While new facilities are well used, there is no evidence of reduction in A&E workload and there is a risk that shorter waiting times might encourage more patients to attend. As I write this in 2004, the work load of A&E departments is increasing faster than in recent years and the reduction of GP involvement in out-of-hours work, due later in the year as a result of a change in their contract, may throw further work on A&E departments.

Implementation of the European Working Time Directive has put further pressures on hospitals by the reduction in junior doctors' hours of work. This has not had a major effect on most A&E departments other than making changes to rotas to ensure that they are compliant with the

Directive. However the effects on hospitals will be significant as it will be impossible to maintain full teams on call in every specialty. This may result in having a single team on call to care for all emergencies in the hospital. The role of nurse practitioners will increase to cope with this. There have been four pilot studies, some of which have included A&E.[4] However there has been discussion about A&E departments (with additional help from inpatient specialties) taking over the care of all emergencies within the hospital at night but care needs to be taken that the staffing, built up over many years, to create a good A&E service is not used to bail out the hospital to the detriment of patients attending A&E.

Clinical changes

Not only has A&E grown since it was recognised as a specialty in the early 1970s but it has also changed considerably as discussed in Chapter 6. With the number of consultants and trainees in post, the specialty is obviously safe, but how will it change over the next few years? A look into the future was given in 1999 by a younger consultant[5] but much has happened since then and although one cannot predict the future, there are certain pointers.

Even in the 1960s the majority of sick patients had medical rather than traumatic problems. The incidence of trauma has reduced considerably since then. The reduced involvement of general practitioners with emergency and out-of-hours work combined with better training and equipping of paramedics means that patients taken ill suddenly are more likely to phone for an ambulance rather than their GP and thus come to A&E. This combined with an aging population means that A&E will become more involved with medical rather than surgical and traumatic problems.

In the past a patient seeking medical help with, for example, chest pain was clinically assessed, investigated with one or more ECGs and possibly blood tests and a decision made on management. The vast majority of decisions made were correct but inevitably a few patients with 'indigestion' later turned out to have a myocardial infarction or unstable angina. There has been a change in attitude over the past few years that clinical judgement (with its occasional, inevitable, error) is inadequate and that acute coronary syndrome needs to be excluded with 100 per cent confidence. Other diseases, too, including deep vein thrombosis, pulmonary embolus, subarachnoid haemorrhage also need to be excluded with 100 per cent confidence. This has led to the concept of observation units where patients will be observed for a few hours and have a protocol-driven series of investigations to exclude the disease in question. Such observation units will often be run by A&E consultants.

For more than 40 years, reports have been recommending 24-hour senior cover in A&E and even 24-hour consultant cover as a way to improve patient care. With increasing numbers of consultants, this may become a reality but this will also mean that the work of consultants will change significantly.

Appendix A: Members Present at the First Committee Meeting of the Casualty Surgeons Association 12 October 1967

M Ellis (chairman)
EP Abson
DB Caro
J Collins
JF Hindle
CD James
D McCarthy
J Pascall
I Stillman
Apologies were received from DM Proctor.

Appendix B: Officers of Casualty Surgeons Association and British Association for Accident and Emergency Medicine

President

Maurice Ellis	1967–72
David Caro	1972–75
Edward Abson	1975–78
John Collins	1978–81
William Rutherford	1981–84
David Wilson	1984–87
David Williams	1987–90
Norman Kirby	1990–93
Keith Little	1993–95
Christopher Cutting	1995–98
Roger Evans	1998–2001
John Heyworth	2001–04
Martin Shalley	2004–

Honorary Secretary

Edward Abson	1967–75
Sheila Christian	1975–78
David Williams	1978–84
John Thurston	1984–90
Stephen Miles	1990–96
John Heyworth	1996–99
Stephen McCabe	1999–

Honorary Treasurer

John Hindle	1967–78
Ian Stewart	1978–84
Stuart Lord	1984–87
Gautam Bodiwala	1987–93
Carlos Perez-Avila	1993–99
Peter Burdett-Smith	1999–

Appendix C: The First Board of the Faculty of Accident and Emergency Medicine 1993

Officers

Dr DJ Williams, President
Maj. Gen. NG Kirby, Vice President
Prof. DW Yates, Dean
Mr GG Bodiwala, Treasurer
Dr JGB Thurstan, Registrar

Elected members

Dr HR Guly
Mr J Marrow
Mr SAD Miles

Nominated members

Dr H Baderman
Dr A McGowan
Mr DV Skinner

Royal College Representatives

Mr IWR Anderson	(RCP&SGlas)
Dr PF Baskett	(RCAnaes)
Prof. NL Browse	(RCSEng)
Prof. IR Cameron	(RCPLond)
Dr JB Irving	(RCPEd)
Mr IB MacLeod	(RCSEd)

Appendix D: Officers of the Faculty of Accident and Emergency Medicine

President

David Williams	1993–96
Keith Little	1996–99
Ian Anderson	1999–2002
Alastair McGowan	2002–

Vice Presidents

Norman Kirby	1993–94
Edward Glucksman	2004–
Jonathan Marrow	2004–

Registrar

John Thurston	1993–97
Edward Glucksman	1997–2004
Ruth Brown	2004–

Dean

David Yates	1993–98
Alastair McGowan	1998–2000
David Skinner	2000–

Treasurer

Gautam Bodiwala	1993–98
James Wardrope	1998–2003
Kevin Reynard	2004–

References

Preface

1. www.acep.org/1.9.0htm (2004).
2. www.acem.org.au/open/documents/history.htm (2004).
3. Anonymous. Accident and emergency services. *Lancet* (1969) 1: 90.
4. Anonymous. A disgraceful situation. *Lancet* (1970) 2: 861.

1 Casualty Staffing before Platt

1. Standing Medical Advisory Committee. *Accident and Emergency Services* (1962) HMSO.
2. Brown L. (Ed). *The New Shorter Oxford English Dictionary Vol 1 (A-M)* (1993) Clarendon Press, Oxford.
3. Ell B. Reflexions of a lexicographer. Casuals and casualties. *BMJ* (1972) 1: 1113.
4. Abson EP. What is 'emergency'? *BMJ* (1980) 280: 1536.
5. Moore N. *The History of St Bartholomew's Hospital Vol II* (1918) C Arthur Pearson, London.
6. Abel-Smith B. *The Hospitals 1800–1948* (1964) Heinemann, London.
7. Fissell ME. *Patients, Power, and the Poor in Eighteenth-Century Bristol* (1991) Cambridge University Press.
8. Clarkson P. Outpatient arrangements and accident services. *Guy's Hospital Gazette* (1948) 31.7.48: 202–11.
9. Power D and Waring HJ. *A Short History of St Bartholomew's Hospital 1123-1923* (1923) St Bartholomew's Hospital, London.
10. Louden ISL. Historical importance of outpatients. *BMJ* (1978) 2: 974–7.
11. Nuffield Provincial Hospitals Trust. *Casualty Services and their Setting* (1960) Oxford University Press.
12. Dale J. GPs in A&E departments in Salisbury C, Dale J and Hallam L (Eds) *24-hour Primary Care* (1999) Radcliffe Medical Press, Abingdon.
13. Clarkson P. Timing in surgery and accident services. *Guy's Hospital Gazette* (1949) 16.7.49: 219–23.
14. Patel AR. Modes of admission to hospital: a survey of emergency admissions to a general medical unit. *BMJ* (1971) 1: 281–30.
15. Catford EF. *The Royal Infirmary of Edinburgh 1929–1979* (1984) Scottish Academic Press, Edinburgh.
16. Lowden TG. The casualty department: the work and the staff. *Lancet* (1956) 1: 955–6.
17. Lowden TG. The casualty department: shortcomings and difficulties. *Lancet* (1956) 1: 1006.
18. Lowden TG. The casualty department: a comprehensive accident service. *Lancet* (1956) 1: 1060–2.

19 Anonymous. Casualty department. *Lancet* (1960) 2: 689–90.
20 Lowden TG. *The Casualty Department* (1955) E&S Livingstone, Edinburgh.
21 Clarkson P. The role of casualty departments in the hospital service. *Guy's Hospital Gazette* (1960) 74: 408–15.
22 Mestitz P. A series of 1817 patients seen in a casualty department. *BMJ* (1957) 2: 1108–9.
23 Caro DB. The Casualty Surgeons Association. *Postgrad Med J* (1972) 48: 260–1.
24 Parry CPB, Cooper BD, Shelswell ME, Dunkerley GE, Denham RA and Murray CMM. The Portsmouth Casualty Service. *BMJ* (1962) 2: 909–12.
25 Black WR. What is an accident? *Lancet* (1959) 2: 124–5.
26 Platt H. British Orthopaedic Association: first founders' lecture. *J Bone Joint Surg (Br)* (1959) 41B: 231–6.
27 Osmond-Clarke H. Half a century of orthopaedic progress in Great Britain. *J Bone Jt Surg (Br)* (1950) 32B: 620–75.
28 Committee on Fractures. Report. *BMJ* (1935) 1(suppl): 53–62.
29 Quoted in BOA. *Memorandum on Accident Services* (1943) BOA.
30 Robb-Smith AHT. *A Short History of the Radcliffe Infirmary* (1970). Church Army Press Oxford for United Oxford Hospitals.
31 Joint Working Party (Chairman Sir Robert Platt). *Medical Staffing Structure in the Hospital Service* (1961) HMSO.
32 Shore E. Medical manpower planning. *Health Trends* (1974) 6: 2–5.
33 Royal Commission on Medical Education 1965–1968. *Report* (Todd Report) (1968) HMSO.
34 Hurst TW, Linfoot CB and Woodhouse TK. Junior medical staffing in hospital: five years of famine. *Lancet* (1953) 1: 1037–8.
35 Wynn WH. Shortage of junior hospital staff. *Lancet* (1952) 2: 125.
36 Anonymous. Shortage of house officers. *Lancet* (1954) 1: 181–2.
37 Anonymous. *Medical Manpower* (1966) Office of Health Economics, London.
38 Forrester RM and Walton JR. The senior registrar. *Lancet* (1953) 1: 793.
39 Shore E. Medical manpower planning. *Health Trends* (1974) 6: 32–5.
40 Clarkson P. Casualty departments in teaching hospitals. *Lancet* (1966) 2: 31–5.
41 Corbishley K. Shortage of casualty officers. *Lancet* (1953) 2: 454.
42 Gee A. Junior hospital staff. *Lancet* (1953) 1: 696–7.
43 Peyman MA. Casualty officer's responsibility. *Lancet* (1953) 1: 545.
44 Smith KAH. Shortage of casualty officers. *Lancet* (1953) 2: 399.
45 CSA/BAEM Archive.
46 Memorandum submitted by SCO Subcommittee 30.10.63.
47 Minutes of SCO Subcommittee 30.10.63.
48 Abson EP and Caro DB. Senior Casualty Officers. *BMJ* (1965) 2: 941.
49 Fry L. Casualties and casuals. *Lancet* (1960) 1: 163–6.
50 Jones PF, Karmody AM and Galloway JMD. A children's casualty department. *BMJ* (1966) 2: 819–21.
51 Tatham PH. Experience with an appointment system in a casualty department. *Lancet* (1966) 1: 1201–3.
52 Garden RS. The casualty department and the accident service. *Lancet* (1965) 1: 901–3.
53 Naylor A. The Bradford Accident Service 1960–1965. *J Roy Coll Surg Edin* (1967) 12: 264–74.

54 Wilson RI and Rutherford WH. The Belfast accident service. *Injury* (1972) 3: 169–75.
55 Hardy RH. Structure and function of a middle-sized accident department. *BMJ* (1974) 2: 596–600.
56 Wainwright D and Steel WM. Accident and emergency services in Stoke-on-Trent. *Injury* (1972) 3: 158–63.
57 Naylor A. The Bradford accident service. *Injury* (1972) 3: 148–57.
58 Denham RA. The accident and emergency service in Portsmouth – present and future. *Postgrad Med J* (1972) 48: 268–74.
59 Clarkson P. Some observations on the casualty department of New Guy's House. *Guy's Hospital Gazette* (1960) 74: 424–30.
60 Heaman EA. *St Mary's. The History of a Teaching Hospital* (2003) McGill-Queens University Press, Montreal and Kingston, pp. 428–9.
61 Howard Baderman, personal communication.
62 Scott JC. Report on the Oxford Accident Service after 25 years. *BMJ* (1967) 2: 632–5.
63 Burrough EJR. *Unity in Diversity: The Short Life of the United Oxford Hospitals* (1978). Privately printed by author.
64 Plewes LW. *Introduction in Accident Service*. Plewes LW (Ed.) (1966) Pitman Medical Publishing Co, London.
65 Anonymous. 'A pretty ghastly, awful picture'. *BMJ* (1961) 2: 1548–9.

2 Who Should Run A&E Departments?

1 British Orthopaedic Association. Memorandum on Accident Services 1959. *J Bone Joint Surg (Br)* (1959) 42B: 458–63.
2 Anonymous. What is an accident? *Lancet* (1959) 2: 75–6.
3 An Accident service for the nation *BMJ* (1959) 2: 1009–10.
4 Fry J. An accident service for the nation. *BMJ* (1959) 2: 1176.
5 Nuffield Provincial Hospitals Trust. *Casualty Services and their Setting* (1960) Oxford University Press.
6 Accident Services Review Committee of Great Britain and Ireland (Chairman H Osmond-Clarke). *Interim Report* (1961) BMA, London.
7 Standing Medical Advisory Committee. *Accident and Emergency Services* (the Platt Report) (1962) HMSO.
8 Rutherford WH. The inter-related problems of an accident service and the accident and emergency department. *Injury* (1975) 7: 96–100.
9 Joint Working Party (Chairman Sir Robert Platt). *Medical Staffing Structure in the Hospital Service* (1961) HMSO.
10 Lamont D. The casualty dilemma. *Lancet* (1961) 2: 1190–3.
11 Ellis M. The casualty department. *Lancet* (1961) 2: 1303.
12 Lawrence W. The casualty department. *Lancet* (1961) 2: 1302.
13 Lamont D. Accident services. *BMJ* (1963) 2: 121.
14 Anonymous. Accident and emergency services. *Lancet* (1962) 2: 544–5.
15 da Costa. GIB. Accident services. *BMJ* (1963) 1: 1742.
16 First meeting of SCO Subcommittee of CCSC 27.6.63.CSA/BAEM Archive.
17 Letter from EP Abson to da Costa 1962. CSA/BAEM Archive.
18 Pascall KG. Senior casualty officers. *Lancet* (1963) 2: 1178.
19 CSA/BAEM Archive.

20 Letter from Mr Abson to Mr da Costa 4.11.63. re. meeting. CSA/BAEM Archive.
21 Ministry of Health. Medical staffing structure in hospitals. *BMJ* (1964) 2: 438–40.
22 Letter from Mr Caro to Mr Walpole Lewin. CSA/BAEM Archive.
23 Orthopaedic Group comments received at CCSC meeting 18.6.64.
24 Casualty response to comments on casualty resolution made by Orthopaedic Group Committee (1964) CSA/BAEM Archive.
25 Ekin WH and Wilson JW. Accident and emergency services. *BMJ* (1965) 2: 1436.
26 Anonymous. Accident and emergency services. *BMJ* (1970) 4: 68.
27 Scott JC. Accident and emergency services. *BMJ* (1965) 2: 1371.
28 Participant at discussion at symposium on the organisation and staffing of the casualty services. *Postgrad Med J* (1972) 48: 276–89.
29 Durbin FC. Casualty departments. *Postgrad Med J* (1972) 48: 262–5.
30 Steel WM. The role of the orthopaedic surgeon in the accident and emergency department. *Injury* (1977) 9: 43–9.
31 O'Connor BT. Conclusions – and a proposed pattern for reorganisation. *Postgrad Med J* (1972) 48: 290–4.
32 Clarkson P. Casualty departments in teaching hospitals. *Lancet* (1966) 2: 31–5.
33 Collins J. The casualty consultant. *BMJ* (1966) 1: 359.
34 Garden RS. The casualty department and the accident service. *Lancet* (1965) 1: 901–3.
35 Letter from Mr Abson to Mr Caro 18.2.65. CSA/BAEM Archive.
36 Letter from E Grey-Turner deputy secretary BMA to Mr Caro 7.7.65. CSA/BAEM Archive.
37 Anonymous. Medical Assistants: Ministry figures. *BMJ* (1968) 4 (suppl): 57–90.
38 Anonymous. *Lancet* (1966) 1: 933.
39 Letter from Dr A Winner (Ministry of Health) to Dr E Grey-Turner (BMA) 6.10.66. CSA/BAEM Archive.
40 Attendance register, CSA/BAEM Archive.
41 Anonymous. A disgraceful situation. *Lancet* (1970) 2: 861.
42 Scott PJ, Durbin FC and Morgan DC. *Accident and Emergency Services in England and Wales in 1969* (1972) BOA. Published also as part of BOA. *Casualty departments: the accident commitment* (1973) BOA, London.
43 Anonymous. Accident and emergency services. *BMJ* (1971) 3: 385–6.
44 Anonymous. Medical students in casualty departments. *BMJ* (1971) 4: 311.
45 Naylor A. The Bradford Accident Service. *Injury* (1972) 3: 148–57.
46 Joint Working Party on the Organisation of Medical Work in Hospitals. *All in a Working Day* (1971) HMSO.
47 Bainbridge JM. Accident and emergency services. *Postgrad Med J* (1972) 48: 254–5.
48 Watts JC Accident Services. *Injury* (1969) 1: 112–14.
49 Ellis M. Staffing of casualty departments. *BMJ* (1972) 4: 608.
50 Ellis M. Accident and emergency services. *BMJ* (1970) 4: 800.
51 Jenkinson J, Moss M and Russell I. *The Royal: The History of Glasgow Royal Infirmary 1794–1994* (1994). HarperCollins, Glasgow.
52 BOA. *Casualty Departments: The Accident Commitment* (1973) BOA, London.
53 Scott JC. Accident services in the 1970s. *Injury* (1970) 2: 1–4.
54 Ekin WH and Wilson JW. Accident and emergency services. *BMJ* (1965) 2: 1436.

55 Waugh W. *A History of the British Orthopaedic Association: The First 75 Years* (1993). British Orthopaedic Association, London.
56 Wainwright D and Steel WM. Accident and Emergency services in Stoke-on-Trent. *Injury* (1972) 3: 158–63.
57 Denham. RA. The accident and emergency service in Portsmouth – present and future. *Postgrad Med J* (1972) 48: 268–74.
58 da Costa GIB. The casualty consultant. *Lancet* (1965) 2: 134.
59 Caro D. Use of resuscitation rooms. *BMJ* (1967) 2: 763.
60 Royal Commission on Medical Education. Report. (Todd Report) (1968) HMSO.
61 Harris NH. Accident and emergency services. *BMJ* (1970) 4: 429.
62 Capener N. Organisation of accident services. *Br J Surg* (1970) 57: 769–70.
63 Turney JP. Staffing of casualty departments. *BMJ* (1972) 4: 607.
64 Norcross K. Personal view. *BMJ* (1972) 4: 421.
65 Cull T. The general practitioner's view. *Postgrad Med J* (1972) 48: 266–7.
66 Charnley J. Unnecessary Xrays. *BMJ* (1971) 2: 278–9.
67 Durbin FC and Batchelor JS. Accident and emergency services. *BMJ* (1971) 3: 432–3.
68 Rigby-Jones G. Accident and emergency services. *BMJ* (1971) 4: 48–9.
69 Abson EP and Caro DB. The casualty consultant. *Lancet* (1965) 1: 1158–9.
70 Elson RA. Problems of staffing of emergency departments. *Postgrad Med J* (1972) 48: 275–7.
71 Anonymous. Accident and emergency services. *Lancet* (1969) 1: 90.
72 DoH HM (68)83 Accident and emergency services (1968).
73 Potter JM. Accident and emergency services *BMJ* (1970) 4: 800–1.
74 Letter from Mr Abson to Mr JC Watts 8.1.67. CSA/BAEM Archive.
75 Duthie RB. The training of an accident surgeon. *Injury* (1971) 2: 279–82.
76 Chatterjee SN Accident and emergency services *BMJ* (1970) 4: 429–30.
77 Turney JP. Staffing of accident and emergency departments *BMJ* (1973) 1: 486.
78 Garden RS. The casualty consultant. *Lancet* (1965) 1: 1333.
79 Bremner AE. Importance of casualty departments. *BMJ* (1970) 4: 113.
80 Bracey DW and Loder RE. Accident and emergency services. *BMJ* (1971) 4: 48.
81 London PS. Accident Services. *Br J Hosp Med* (1978) 20: 169–77.
82 Committee (Chairman Sir Max Rosenheim). *Medical Work in Accident and Emergency (Casualty) Departments* (1969). Royal College of Physicians, London.
83 Baird RN, Noble J and McLean D. Initial intensive care in an accident and emergency department. *BMJ* (1972) 4: 90–2.
84 Working Party. *Career Structure in Accident and Emergency Departments* (1971) CCHMS.
85 Anonymous. Staffing in accident departments. *BMJ* (1971) 2(suppl): 13–14.
86 Anonymous. Accident departments. *BMJ* (1971) 3(suppl): 62.
87 JCC 26.1.71.
88 JCC Subcommittee 26.3.71.
89 Letter from Secretary RCSEng to Sir John Richardson Chairman JCC 23.3.71.
90 Letter from Henry Yellowlees to Senior Administrative Medical Officers, Regional Hospital Boards 12.11.71.

3 The First Consultants

1. Wilson DH, personal communication.
2. Ellis M. Outpatient treatment of the injured hand. *Lancet* (1951) 1: 1038–41.
3. Ellis M. The use of penicillin and sulphonamides in the treatment of suppuration. *Lancet* (1951) 1: 774–5.
4. Ellis M. Tenosynovitis of the wrist. *BMJ* (1951) 2: 777–9.
5. Ellis M. Accident and emergency services. *BMJ* (1970) 4: 800.
6. Baderman H, personal communication.
7. Swann I, personal communication.
8. Matheson AB. Obituaries: David Proctor *BMJ* (1994) 309: 1013–14.
9. Matheson AB, personal communication.
10. Anonymous. Staffing in accident departments. *BMJ* (1971) 2(suppl): 13–14.
11. Durbin FC and Batchelor JS. Accident and emergency services. *BMJ* (1971) 3: 432–3.
12. Durbin FC Casualty departments. *Postgrad Med J* (1972) 48: 262–5.
13. Elson R. The medical staff of British accident and emergency units. *Br J Hosp Med* (1971): 161–70.
14. Review of 'pilot' scheme by DHSS 1974, published as appendix C in Lewin W. *Medical Staffing of Accident and Emergency Services*. JCC 1978.
15. Lewin W. *Medical Staffing of Accident and Emergency Services*. JCC 1978.
16. Letter from Henry Yellowlees to Senior Administrative Medical Officers, Regional Hospital Boards 12.11.71.
17. Stewart I, Personal communication.
18. Dickenson J and Shand W. David Bernard Caro: Obituary. *BMJ* (1996) 313: 164.
19. McKie D. Accident and emergency services. *Lancet* (1973) 2: 1250.
20. CSA Comm 3.10.72.
21. Williams HO. Consultants in A&E. *Lancet* (1972) 2: 761.
22. Potter JM. Consultants in A&E. *Lancet* (1972) 2: 923–4.
23. Cutting C, personal communication.
24. Anonymous. Staffing accident and emergency services. *BMJ* (1979) 1: 1363.
25. Anonymous. The improving image of A and E. *BMJ* (1979) 2: 1314.
26. Pritty P. Accepting accident and emergency medicine as a specialty. *BMJ* (1981) 282: 1324.
27. Medical Staffing Division, DHSS. Hospital Medical Staffing in the National Health Service in England and Wales. *Health Trends* (1977) 8: 45–7.
28. BOA. *Casualty Departments: The Accident Commitment* (1973) BOA, London.
29. Joint Working Party of JCHMT and JCHST 1975.
30. Anonymous. Accident and emergency departments. *BMJ* (1977) 2: 123.
31. Anonymous. Appointment of accident and emergency consultants. *BMJ* (1980) 281: 467.
32. London PS. Accident Services. *Br J Hosp Med* (1978) 20: 169–77.
33. James JIP. Consultants in accident and emergency medicine. Paper produced for the BOA (1978).
34. CSA Comm 3.5.78.
35. Caro D in discussion at symposium on the organisation and staffing of the casualty services. *Postgrad Med J* (1972) 48: 276–289.

36 Operational policy and proposed schedule of duties for consultant (1972) Accident Department, Taunton.
37 CSA Comm 15.10.74.
38 Working Party of the JCC/GMSC. *The Staffing of Accident and Emergency Departments* (Mills Report) (1981) JCC/GMSC.
39 Cain D, personal communication.
40 Ahmed O. Accident and emergency services. *BMJ* (1980) 280: 119.
41 Morgan WJ. Accident and emergency services. *BMJ* (1979) 2: 1590.
42 Waugh W. *A History of the British Orthopaedic Association: The First 75 Years* (1993) British Orthopaedic Association, London.
43 Steel WM. The role of the orthopaedic surgeon in the accident and emergency department. *Injury* (1977) 9: 43–9.
44 Dove A, personal communication.
45 James JIP. British Orthopaedic Association Presidential Address. *J Bone Joint Surg (Br)* (1978) 60B: 131–5.
46 Minutes of a meeting at DHSS 22.9.71 between JCC and DHSS to discuss consultants in charge of emergency services.
47 Anonymous. Accident and emergency departments. *BMJ* (1978) 2: 1314.
48 Wilson DH. The multi-consultant accident and emergency department. *Br J Accid Emerg Med* (1983) 1(4): 4–6.
49 Letter from Dr Guly to Dr Dallos (Chairman of the CSA Clinical Services Committee) 3.6.82.

4   Senior Registrars and Training

1 Lewin W. *Medical Staffing of Accident and Emergency Services* (1978) JCC.
2 Letter from sec RCSEng to Sir John Richardson Chairman JCC 23.3.71.
3 Wilson D. The development of accident and emergency medicine. *Community Medicine* (1980)2: 28–35.
4 CSA Comm. 9.10.73.
5 CSA Comm. 3.10.72.
6 Casualty Surgeons Association. *An Integrated Emergency Service* (1973) Casualty Surgeons Association.
7 CSA Comm. 14.2.73.
8 Joint Working Party of JCHMT and JCHST Oct. 74–Jan. 75.
9 Wilson D, personal communication.
10 Jenkinson J, Moss M and Russell I. *The Royal: The History of the Glasgow Royal Infirmary 1794–1994*. Glasgow Royal Infirmary NHS Trust.
11 JCHMT. 2 Report (1975) JCHMT.
12 Tate M. Training in accident and emergency medicine. *Health Trends* (1976) 8: 79–80.
13 Baderman H, personal communication.
14 Matheson AB, personal communication.
15 Lord S, personal communication.
16 Ferguson D, personal communication.
17 Rutherford W, personal communication.
18 Working Party of the JCC/GMSC. *The Staffing of Accident and Emergency Departments* (Mills Report) (1981) JCC/GMSC.

19 Adams I, Flowers MW, Gosnold JK, Marsden AK and Wilson DH. The MRCGP exam and accident and emergency departments. *BMJ* (1983) 287: 361.
20 Medical Staffing Division, DHSS. Hospital Medical Staffing in the National Health Service in England and Wales. *Health Trends* (1982) 14: 28–33.
21 Medical Staffing Division, DHSS. Hospital Medical Staffing in the National Health Service in England and Wales. *Health Trends* (1983) 15: 35–9.
22 Miles S. The senior registrar training programme. *Br J Accid Emerg Med* (1983) 1(3): 4–5.
23 Notes of a meeting between Profs Davidson and Miles (Postgraduate Deans of Universities of Liverpool and Manchester) with Mr M Hall and G Laing. Nov. 1974. CSA/BAEM Archive.
24 CSA Ex. 16.10.81.
25 A&E SRs Travelling Club minutes Jan/Jun/Oct 1981.
26 CSA Ex. 16.7.82.
27 Bowers DM. Staffing of accident and emergency departments. *BMJ* (1978) 2: 1648.
28 CSA Ex. 21.1.83.
29 Driscoll P, Cope A and Miles SAD. Adequacy of senior registrar training in accident and emergency medicine over the last 5 years. *Arch Emerg Med* (1988) 5: 162–8.
30 Johnson G, Brown R and Howell M. Higher specialist training in accident and emergency medicine – past, present and future. *J Accid Emerg Med* (1997) 14: 104–6.
31 BAEM Ex. 4.7.91.
32 Wyatt JP. The role of the accident and emergency registrar. *J Roy Soc Med* (1994) 87: 697–700.
33 BAEM Ex. 8.4.91.
34 Department of Health. *A Guide to Specialist Registrar Training* (1996) NHS Executive.

5 How Many Consultants?

1 Casualty Surgeons Association. *An Integrated Emergency Service* (1973) Casualty Surgeons Association, London.
2 CSA. *Medical Staffing Norms for A&E Departments* (1980) CSA.
3 Lewin W. *Medical Staffing of Accident and Emergency Services* (1978) JCC.
4 Anonymous. How should accident and emergency departments be run? *BMJ* (1979) 2: 1051–3.
5 Working Party of the JCC/GMSC. *The Staffing of Accident and Emergency Departments (Mills Report)* (1981) JCC/GMSC.
6 Anonymous. A&E departments. *BMJ* (1981) 283: 446–7.
7 Appleyard WJ. Medical manpower mismanagement. *BMJ* (1982) 284: 1351–5.
8 Social Services Committee. *Medical Education. Vol 1* (Short Report) (1981) HMSO.
9 CSA Ex. 29.1.82.
10 CSA Working Party. *Discussion Paper on Short Report* (1982).
11 Letter from Dr Guly to Dr Dallos 30.11.82 giving SR response to working party paper.

12. Letter from Dr Guly to Dr Dallos 3.6.82.
13. CSA AGM. 28.5.80.
14. CSA Ex. 4.3.83.
15. CSA Ex. 21.1.83.
16. CSA AGM. 13.4.83.
17. Rutherford WH. Organisation of trauma services in Britain: the future direction. *Arch Emerg Med* 1984: 1 supplement: 53–6.
18. Rutherford W, personal communication.
19. CSA Comm. 30.9.77.
20. CSA AGM. 26.4.84.
21. CSA Ex. 18.1.85.
22. A&E subcommittee of CCHMS 7.9.85.
23. CSA Ex. 10.1.86.
24. CSA Ex. 8.10.87.
25. CSA Ex. 4.7.86.
26. CSA Ex. 5.1.89.
27. Harrison SH. Accident surgery – the life and times of William Gissane. *Injury* (1984): 145–54.
28. CSA AGM. 18.5.85.
29. CSA Ex. 3.4.89.
30. Mason MA. Editorial. *Br J Accid Emerg Med* (1987) 2 (1): 3.
31. Anonymous. OME survey confirms consultants' commitment to NHS *BMJ* (1990) 300: 471–2.
32. National Audit Office. *NHS Accident and Emergency Departments in England* (1992) HMSO.
33. UK Health Departments, JCC and Chairmen of the RHAs. Hospital Medical Staffing – Achieving a Balance. *BMJ* (1986) 293: 147–51.
34. Commission on the Provision of Surgical Services Working Party. *The Management of Patients with Major Injuries* (1988) Royal College of Surgeons of England.
35. Dowie R. *Patterns of Hospital Medical Staffing: Overview* (1991) HMSO.
36. NHS Management Executive. *Hour of Work of Doctors in Training: The New Deal* (1991) EL (91) 82.
37. Allen P. Medical and dental staffing prospects in the NHS in England and Wales 1992. *Health Trends* (1993) 25: 118–26.
38. National Audit Office. *NHS Accident and Emergency Departments in England* (1992) HMSO.
39. BAEM Ex. 9.1.92.
40. BAEM AGM. 2.4.93.
41. FAEM board 15.12.94, Paper from JCHMT numbers of SRs accredited.
42. Working Party on Future Medical Staffing BAEM Ex. 7.10.93.
43. Report on joint meeting of BAEM and FAEM to discuss problems with SHO recruitment, FAEM Board 21.9.95.
44. BAEM AGM. 29.3.96.
45. Department of Health. *A Guide to Specialist Registrar Training* (1996) NHS Executive.
46. Paper circulated at FAEM AGM 2000.
47. BAEM Council 2.7.98.
48. FAEM Board 9.3.99.

49 Letter from NHSE to FAEM Board 16.9.99.
50 BAEM and FAEM. *Workforce Planning in A&E Medicine 2001–2010* (July 2001).

6 A Changing Specialty

1. Nuffield Provincial Hospitals Trust. *Casualty Services and their Setting* (1960) Oxford University Press.
2. Durbin FC. Report of Accident Services Committee of the British Orthopaedic Association *J Bone Jt Surg (Br)* (1971) 53B: 765–6.
3. Commission on the Provision of Surgical Services Working Party. *The Management of Patients with Major Injuries* (1988) Royal College of Surgeons of England.
4. BAEM Ex. 12.7.90.
5. Information extracted from BAEM Directory 1996. BAEM, London.
6. Lowden TG. The casualty department: the work and the staff. *Lancet* (1956) 1: 955–6.
7. Standing Medical Advisory Committee. *Accident and Emergency Services* (the Platt Report) (1962) HMSO.
8. Skinner D. Accident and emergency services. *BMJ* (1990) 301: 1292.
9. Cooke MW, Kelly C, Khattab A, Lendrum K, Morrell R and Rubython EJ. Accident and emergency 24-hour senior cover – a necessity or a luxury. *J Accid Emerg Med* (1998) 15: 181–4.
10. Wyatt JP, Henry J and Beard D. The association between the seniority of accident and emergency doctor and outcome following trauma. *Injury* (1999) 30: 165–8.
11. Binchy J. Accident and emergency medicine – the next 25 years. *J Accid Emerg Med* (1999) 16: 48–54.
12. Nicholl J and Turner J. Effectiveness of a regional trauma system in reducing mortality from major trauma; before and after study. *BMJ* (1997) 315: 1349–54.
13. Prescott M, personal communication.
14. Yates D. Regional trauma systems. *BMJ* (1997) 315: 1321–2.
15. Miles S, personal communication.
16. McCabe S, personal communication.
17. Cooke M, personal communication.
18. BAEM AGM. 29.3.96.
19. BAEM and FAEM. Workforce planning in A&E Medicine 2001–2010 (2001).
20. Social Services Committee. *Medical Education. Vol. 1* (Short Report) (1981) HMSO.
21. Working Party. *SHO Training: Tackling the Issues, Raising the Standards* (1995) Committee of Postgraduate Medical Deans (COPMED) and UK Conference of Postgraduate Deans.
22. Department of Health. *The NHS Plan* (2000) Department of Health.
23. Yates DW. Accident and emergency services. *BMJ* (1991) 302: 111.
24. Lecky F, Woodford M and Yates DW. Trends in trauma care England and Wales 1989–97. *Lancet* (2000) 355: 1771–5.
25. Heyworth J. Medical staffing in A&E in Coffey T and Mythen M (eds) *NHS Frontline – Visions for 2010* (2000) The New Health Network, London.

26 Anonymous. OME survey confirms consultants' commitment to NHS. *BMJ* (1990) 300: 471–2.
27 Brown R. Activities of accident and emergency consultants – a time and motion study. *J Accid Emerg Med* (2000): 17122–5.
28 BAEM and FAEM. Report of Joint Meeting on extension of role of A&E consultant 29.6.99.
29 Report of the Working party on the Management of Head Injuries. (1999) Royal College of Surgeons of England.
30 Jackson RH. Children in accident and emergency departments. *BMJ* (1985) 291: 991–2.
31 Mason MA. Children in accident and emergency departments. *BMJ* (1985) 291: 1353.
32 CSA Ex. 8.10.87.
33 British Paediatric Association, British Association of Paediatric Surgeons and Casualty Surgeons Association. *Joint Statement on Children's Attendances at Accident and Emergency Departments* (1988) British Paediatric Association.
34 Multidisciplinary Working Party. *Accident and Emergency Service for Children* (1999) Royal College of Paediatrics and Child Health.
35 Marsden A, personal communication.
36 Walker A and Brenchley J. Survey of the use of rapid sequence induction in the accident and emergency department. *J Accid Emerg Med* (2000) 17: 95–7.
37 Nee P, personal communication.
38 Baird RN, Noble J and McLean D. Initial intensive care in an accident and emergency department. *BMJ* (1972) 4: 90–2.
39 Bache JB. The work of an A&E department: a new look at the figures. *A&E News* (1982) March 4–5.
40 Shalley MJ and Cross AB. Which patients are likely to die in an accident and emergency department? *BMJ* (1984) 289: 419–21.
41 Anonymous. Rising emergency admissions. *BMJ* (1995) 310: 207–8.
42 Anonymous. The continuing rise in emergency admissions. *BMJ* (1996) 312: 991–2.
43 Volans A. Trends in emergency admissions. *BMJ* (1999) 319: 1201.
44 Penny WJ. Improving door-to-needle times for thrombolysis in acute myocardial infarction. *J Roy Coll Phys Lond* (1999) 33: 6–7.
45 Working Party of the Federation of Medical Royal Colleges. *Acute Medicine: The Physician's Role* (2000) Royal Colleges of Physicians.
46 Quin G. Chest pain evaluation units *J Accid Emerg Med* (2000) 17: 237–40.
47 Mather HM and Connor H. Coping with pressures in acute medicine – the second RCP consultant questionnaire survey. *J Roy Coll Phys Lond* (2000) 34: 371–3.
48 Schiller KFR. Specialists should not be expected to practise general medicine. *BMJ* (1999) 318: 1759.
49 Working Party. Acute medicine: making it work for patients. A blueprint for organisation and training (2004) Royal College of Physicians, London.

7 Academic A&E, the Faculty and Changes of Name

1 Clarkson P. The role of casualty departments in the hospital service. *Guys Hospital Gazette* (1960) 74: 408–15.

2   Wilson DH. *Aims and Objectives of Teaching in Accident and Emergency*; ASME Occasional Publication 1 (1981) Association for the Study of Medical Education, Dundee.
3   Lowden TG. *The Casualty Department* (1955) E&S Livingstone Ltd Edinburgh.
4   Ellis M. The use of penicillin and sulphonamides in the treatment of suppuration. *Lancet* (1951) 1: 774–5.
5   Ellis M. Tenosynovitis of the wrist. *BMJ* (1951) 2: 777–9.
6   CSA 1st meeting 12.10.67.
7   Foex BA, Dark PM and Yates DW. On the retirement of Professor Rod Little *EMJ* (2003) 20: 2.
8   Irving M. Emergency Care – an academic specialty. *Resuscitation* (1977) 5: 197–204.
9   BAEM Ex. 8.7.93.
10 Rutherford WH. The medical effects of seat-belt legislation in the United Kingdom. *Arch Emerg Med* (1985) 2: 221–3.
11 Emergency Medicine Research Society Members Handbook (1992) Emergency Medicine Research Society.
12 CSA Ex. 12.10.84.
13 CSA Ex. 5.7.85.
14 CSA Ex. 8.10.87.
15 CSA Ex. 16.1.81.
16 CSA Ex. 8.3.90.
17 BAEM Ex. 8.4.91.
18 Notes of a meeting between Profs Davidson and Miles (Postgraduate Deans of Universities of Liverpool and Manchester) with Mr M Hall and G Laing. Nov. 1974 CSA/BAEM Archive.
19 CSA AGM. 3.4.75.
20 CSA AGM. 22.4.76.
21 Letter from R Adams and J McNae to S Christian. 5.12.77 CSA/BAEM Archive.
22 Lewin W. *Medical Staffing of Accident and Emergency Services*. JCC 1978.
23 CSA AGM. 28.5.80.
24 CSA Ex. 6.10.88.
25 CSA Ex. 4.1.90.
26 BAEM Ex. 4.10.90.
27 BAEM Ex. 14.5.90.
28 BAEM Ex. 8.4.91.
29 BAEM Ex. 4.7.91.
30 BAEM Ex. 7.4.92.
31 FAEM Board 15.9.94.
32 Thurston J. How to acquire a coat of arms. *BMJ* (1997) 315: 1682–4.
33 Working Group on Specialist Medical Training. *Hospital Doctors: Training for the Future* (Calman Report) (1993) Health Publications Unit, Heywood, Lancs.
34 Department of Health. *A Guide to Specialist Registrar Training* (1996) NHS Executive.
35 FAEM Board 2.3.00.
36 FAEM Board 17.9.98.
37 FAEM Board 17.12.98.
38 Ryan JM and Heyworth J. Casualty is outdated term for emergency medicine. *BMJ* (2002) 324: 422.

39 Letter from HH Langston to Mr Abson. 27.11.67 CSA/BAEM Archive.
40 CSA Comm. 1.4.72.
41 Mason MA. Editorial. *Br J Accid Emerg Med* (1986) 1 (4): 3.
42 Rocke LG. Accident and Emergency or emergency medicine. *J Accid Emerg Med* (1999) 16: 74.
43 Reid C. Role of accident and emergency doctors should be expanded. *BMJ* (2000) 320: 1728.
44 Davis RM and Pless B. *BMJ* bans 'accidents'. *BMJ* (2001) 322: 1320–1.
45 Sakr M and Wardrope J. Casualty, accident and emergency, or emergency medicine, the evolution. *J Accid Emerg Med* (2000) 17: 314–19.
46 FAEM Board 17.6.04.

8 Non-consultant and Non-training-grade Doctors

1 Joint Working Party (Chairman Sir Robert Platt). *Medical Staffing Structure in the Hospital Service* (1961) HMSO.
2 Standing Medical Advisory Committee. *Accident and Emergency Services* (the Platt Report). (1962) HMSO.
3 Royal Commission on Medical Education 1965–1968. *Report* (Todd Report). (1968) HMSO.
4 Working Party. *The Responsibilities of the Consultant Grade* (1969) HMSO.
5 Lewin W. *Medical Staffing of Accident and Emergency Services*. JCC (1978).
6 JCHTA&E Minutes 16.2.99.
7 Anonymous. Reception of casualties. *Br J Surg* (1964) 51: 791–2.
8 Joint Working Party on the Organisation of medical work in Hospitals. *All in a Working Day* (1971) HMSO.
9 Working Party of the JCC/GMSC. *The Staffing of Accident and Emergency Departments* (Mills Report) (1981) JCC/GMSC.
10 Social Services Committee. *Medical Education. Vol 1* (Short Report) (1981) HMSO.
11 CSA Ex. 24.7.81.
12 Marrow J. Emergency in emergency departments. *BMJ* (1977) 2: 1545.
13 UK Health Departments, JCC and Chairmen of the RHAs. *Hospital Medical Staffing – Achieving a Balance – Plan for Action* (1987) DoH.
14 CSA Ex. 4.3.83.
15 CSA Ex. 18.1.85.
16 CSA Ex. 5.7.85.
17 Anonymous. Clinical assistant advertisements. *BMJ* (1984) 288: 1176.
18 Paper produced for Joint Negotiating Committee for Hospital Medical and Dental Staff Working Party meeting Feb. 1985.
19 Dowie R. *Patterns of Hospital Medical Staffing: Overview* (1991) HMSO.
20 Department of Health. *Reforming Emergency Care* (2001) Department of Health, London.
21 Hill P and Donaldson L. Appointment as a staff grade doctor: success of failure? *Health Trends* (1993) 25: 109–11.
22 SCOPME. *Meeting the Educational Needs of Staff Grade Doctors and Dentists* (1994) SCOPME.
23 FAEM Board 13.3.97.

24 BAEM Ex. 8.4.91.
25 BAEM AGM. 29.3.96.
26 BAEM AGM. 28.5.95.
27 BAEM Ex. 29.3.93.
28 21.12.95. Notes of seminar on maintaining A&E services and shortage of SHOs at DoH.
29 Anonymous Accident departments can appoint more staff. *BMJ* (1996) 312: 188.
30 BAEM Policy on Non-Consultant Career Grade Doctors 2002.
31 *BMJ*. Classified 9.12.00.
32 Dosani S, Schroter S, MacDonald R and Connor J. Recruitment of doctors to non-standard grades in the NHS: analysis of job advertisements and survey of advertisers. *BMJ* (2003) 327: 961–4.
33 NHS Executive. *A Health Service for all the Talents: Developing the NHS Workforce 2000.* DoH.
34 DoH. *The NHS Plan* (2000) DoH.

9 Junior Staffing of A&E Departments

1 Nicholas JJ. Shortage of casualty officers. *BMJ* (1970) 4: 113.
2 Guha-Ray DK. Shortage of casualty officers. *BMJ* (1970) 3: 774.
3 Clarkson P. Casualty departments in teaching hospitals. *Lancet* (1966) 2: 31–5.
4 Corbishley K. Shortage of casualty officers. *Lancet* (1953) 2: 454.
5 Standing Medical Advisory Committee. *Accident and Emergency Services* (The Platt report) (1962) HMSO.
6 Garden RS. The casualty department and the accident service. *Lancet* (1965) 1: 901–3.
7 Anonymous. Reception of casualties. *Br J Surg* (1964) 51: 791–2.
8 Joint Working Party (Chairman Sir Robert Platt). *Medical Staffing Structure in the Hospital Service* (1961) HMSO.
9 Anonymous. Memorandum on the current problems of hospital medical staff. *BMJ* (1967) 2(suppl): 93–6.
10 DoH HM (68) 83 *Accident and Emergency Services* (1968).
11 Casualty Surgeons Association. *An Integrated Emergency Service* (1973) Casualty Surgeons Association, London.
12 Swann I, personal communication.
13 Ferguson D, personal communication.
14 Cox P, personal communication.
15 BAEM. *Directory* (1996) BAEM.
16 Elson R. The medical staff of British accident and emergency units. *Br J Hosp Med* (1971): 161–70.
17 Bowers DM. Staffing of accident and emergency departments. *BMJ* (1978) 2: 1648.
18 Slack CC. Emergency in emergency departments. *BMJ* (1977) 2: 1359.
19 CSA AGM.1987.
20 Stewart I. Staffing of accident and emergency departments. *J Accid Emerg Med* (1996) 13: 412–14.

21 BAEM Council 12.1.95.
22 BAEM Council 6.7.95.
23 Minutes of a meeting held on 14.7.95 between BAEM and FAEM to discuss problems in the recruitment of SHOs in A&E.
24 UK Health Departments, JCC and Chairmen of the RHAs. *Hospital Medical Staffing – Achieving a Balance – Plan for Action* (1987) DoH.
25 NHS Management Executive. Hours of work of doctors in training: the New Deal (1991) *EL* (91) 82.
26 Allen P. Medical and dental staffing prospects in the NHS in England and Wales 1992. *Health Trends* (1993) 25: 118–26.
27 Gosnold J. Emergency in emergency departments. *BMJ* (1977) 2: 1673.
28 Royal Commission on Medical Education 1965–1968. *Report* (Todd Report). (1968) HMSO.
29 Medical Workforce Standing Advisory Committee third report. *Planning the Medical Workforce* (1997) DoH.
30 CSA Ex. 18.1.85.
31 CSA Ex. 14.4.85.
32 BAEM Ex. 14.4.94.
33 BAEM Council 10.7.97.
34 BAEM Council 2.7.98.
35 Anonymous. Accident and emergency services: the staffing of departments. *BMJ* (1979): 1119–21.
36 CSA Ex. 16.7.87.
37 BAEM AGM. 1995.
38 CSA Ex. 6.10.88.
39 Anonymous. Hospital medical staffing *BMJ* (1966) 2(suppl): 56.
40 Rigby-Jones G. Casualty staffing. *BMJ* (1967) 4: 360.
41 Anonymous. Accident and emergency departments. *BMJ* (1971) 4(suppl): 81–2.
42 CSA Ex. 24.4.84.
43 CSA AGM. 18.5.85.
44 CSA AGM. 14.4.88.
45 Lewin W. *Medical Staffing of Accident and Emergency Services* (1978) JCC.
46 Tye CC, Ross F and Kerry SM. Emergency nurse practitioner services in major accident and emergency departments; a United Kingdom postal survey. *J Accid Emerg Med* (1998) 15: 31–34.
47 NHS Executive. *A Health Service for all the Talents: Developing the NHS Workforce* (2000) NHSE.
48 Notes of seminar on maintaining A&E services and shortage of SHOs at DoH. 21.12.95.
49 Clinical Services Committee. *Dealing with Shortages of Medical Staff* (1996) BAEM.
50 Lowden TG. The casualty department: the work and the staff. *Lancet* (1956) 1: 955–6.
51 Accident Services Review Committee of Great Britain and Ireland. Report of a Pilot Study conducted by a Working Party on Progress in the Provision of Accident Services (1970) BMA.
52 Cooke MW, Higgins J and Bridge P. *A&E: The Present State* (2000) Emergency Medicine Research Group, Universities of Warwick and Birmingham.

53 Scott PJ, Durbin FC and Morgan DC. *Accident and Emergency Services in England and Wales in 1969* (1972) BOA. Published also as part of BOA. *Casualty Departments: The Accident Commitment* (1973) BOA.
54 Department of Health for Scotland. *Hospital Planning Notes: Provision and Design of Casualty and Accident Departments* (1961) HMSO, Edinburgh.
55 Joint Working Party on the Organisation of medical work in Hospitals. *All in a Working Day* (1971) HMSO.
56 Tham KY, Richmond PW and Evans RJ. Senior house officers' work activities in an accident and emergency department. *J Accid Emerg Med* (1995) 12: 266–9.
57 BAEM. *The Way Ahead* (1997) BAEM.

10 Primary Care in A&E

1 Clarkson P. Timing in surgery and accident services. *Guy's Hospital Gazette* (1949) 16.7.49: 219–23.
2 Dale J. GPs in A&E departments in Salisbury C, Dale J and Hallam L (Eds) *24-hour Primary Care* (1999) Radcliffe Medical Press, Abingdon.
3 Fry L. Casualties and casuals. *Lancet* (1960) 1: 163–6.
4 Williams B, Nicholl J and Brazier J. *Health Care Needs Assessment: Accident and Emergency Departments* (1996) Wessex Institute of Public Health Medicine, Winchester.
5 Leydon GM, Lawrenson R, Meakin R and Roberts JA. The cost of alternative models of care for primary care patients attending accident and emergency departments: a systematic review. *J Accid Emerg Med* (1998) 15: 77–83.
6 Crombie DL. A casualty survey. *J Coll Gen Pract* (1958) 2: 346–56.
7 Blackwell B. Why do patients come to a casualty department. *Lancet* (1962) 1: 369–71.
8 Griffiths WAD, King PA and Preston BJ. Casualty department – or GP service. *BMJ* (1967) 3: 46.
9 Wilkinson A, Kazantzis G, Williams DJ, Dewar RAD, Bristow KM and Miller DL. Attendance at a London casualty department. *J Roy Coll Gen Pract* (1977) 27: 727–33.
10 Davison AG, Hildrey ACC and Floyer MA. Use and abuse of an accident and emergency department in East London. *J Roy Soc Med* (1983) 76: 37–40.
11 Singh S. Self referral to accident and emergency department: patients' perceptions. *BMJ* (1988) 297: 1179–80.
12 Ellis M. Coping with minor casualties. *BMJ* (1974) 2: 55.
13 Davies JOF & Lewin W. Observations on hospital planning. *BMJ* (1960) 2: 763–8.
14 Nuffield Provincial Hospitals Trust. *Casualty Services and their Setting* (1960) Oxford University Press.
15 Standing Medical Advisory Committee. *Accident and Emergency Services* (1962) HMSO.
16 Garden RS. The casualty department and the accident service. *Lancet* (1965) 1: 901–3.
17 Murley RS. Casualty department – or GP service. *BMJ* (1967) 3: 245.
18 Bell AD. Casuals and casualties. *Lancet* (1960) 1: 277–8.

19 DoH HM (68)83 *Accident and Emergency Services*. (1968) DoH.
20 Royal Commission on Medical Education 1965–1968. *Report* (Todd Report). (1968) HMSO.
21 Orthopaedic Group Committee comments on SCO memorandum received at CCSC meeting 18.6.64.
22 Anonymous. Accident and emergency services. *Lancet* (1962) 2: 544–5.
23 Lewin W. *Medical Staffing of Accident and Emergency Services* (1978) JCC.
24 Jones CS and McGowan A. Self referral to an accident and emergency department for another opinion. *BMJ* (1989) 298: 859–62.
25 Nguyen-Van-Tam JS and Baker DM. General practice and accident and emergency department care: does the patient know best? *BMJ* (1992) 305: 157–8.
26 Corbally E. Responsibilities for casualties: medicine and the law. *Lancet* (1967) 2: 1142–3.
27 Ellis M. The casualty department. *Lancet* (1961) 2: 1303.
28 Robson AO. Why patients come to a casualty department. *Lancet* (1962) 1: 539.
29 Jennings JH. Casualty department – or GP service? *BMJ* (1967) 3: 561.
30 Rutherford WH. Accident and emergency medicine in Rutherford WM, Illingworth RN, Marsden AK, Nelson PG, Redmond AD and Wilson DH (Eds) *Accident and Emergency Medicine*. 2nd edition (1989) Churchill Livingstone, Edinburgh.
31 London PS. Accident Services. *Br J Hosp Med* (1978) 20: 169–77.
32 Joint Working Party on the Organisation of medical work in Hospitals. *All in a Working Day* (1971) HMSO.
33 Hardy RH. Structure and function of a middle-sized accident department. *BMJ* (1974) 2: 596–600.
34 Anonymous. Symposium accident and casualty services. *BMJ* (1962) 2: 392–4.
35 Cull T. The general practitioner's view. *Postgrad Med J*. (1972) 48: 266–7.
36 Tomlinson B. *Report of the Enquiry into London's Health Service, Medical Education and Research* (1992) HMSO.
37 Anonymous. Primary care in inner London: inadequate and exposed. *BMJ* (1981) 282: 1739–40.
38 Morrell D. Introduction and overview in Louden I, Horder J and Webster C (Eds) *General Practice under the National Health Service 1948–1977* (1998) Clarendon Press, London.
39 Watson J, Robinson ET, Hardern KA and Crawford RL. Changes in demand for initial medical care in general practice and hospital accident and emergency departments. *BMJ* (1979) 2: 365.
40 Anonymous. Coping with minor casualties. *BMJ* (1974) 1: 339.
41 Reeves AJ. Coping with minor casualties. *BMJ* (1974) 2: 122.
42 Evans EO, Coigley M, Lewis JVV *et al.* Coping with minor casualties. *BMJ* (1974) 2: 440.
43 Scott JC. Coping with minor casualties. *BMJ* (1974) 1: 573.
44 Committee on Child Health Services (Chairman Prof. S Court). *Fit for the Future* (1976) HMSO.
45 Various speakers. Discussion at symposium on the organisation and staffing of the casualty services. *Postgrad Med J* (1972) 48: 276–89.

46 Royal Commission on the National Health Service (chairman Sir Alec Merrison) *Report* (1979) HMSO.
47 Working Party of the JCC/GMSC. *The Staffing of Accident and Emergency Departments* (Mills Report) (1981) JCC/GMSC.
48 Response to Mills Report, Reported at CSA AGM. 22.4.82.
49 Blythin P. Triage – first things first. *Br J Accid Emerg Med* (1983) 1 (3): 15.
50 Joshi M and Parmar M. Letter. *Br J Accid Emerg Med* (1983) 1 (5): 16.
51 Carew-McColl M and Buckles E. A workload shared. *Health Services Journal* (1990) 100: 27.
52 Carew-McColl M. Gently adjusting open doors. *Arch Emerg Med* (1990) 7: 59–60.
53 Dale J, Green J, Reid F, Glucksman E ands Higgs R. Primary care in the accident and emergency department: 11. Comparison of general practitioners and hospital doctors. *BMJ* (1995) 311: 427–30.
54 Dale J, Lang H, Roberts JA, Green J and Glucksman E. Cost effectiveness of treating primary care patients in accident and emergency: a comparison between general practitioners, senior house officers and registrars. *BMJ* (1996) 312: 1340–4.
55 Cooke M. Better to increase number of consultants in accident and emergency medicine. *BMJ* (1996) 313: 628.
56 Ward P, Huddy J, Hargreaves S, Touquet R, Hurley J and Fothergill J. Primary care in London: an evaluation of general practitioners working in an inner city accident and emergency department. *J Accid Emerg Med* (1996) 13: 11–15.
57 Freeman GK, Meakin RP, Laurenson RA, Leydon GM and Craig G. Primary care units in A&E departments in North Thames in the 1990s: initial experience and future implications. *Br J Gen Pract* (1999) 49: 107–10.
58 Fisher J, personal communication.
59 Robertson Steele I, personal communication.
60 Hallam L and Henthorne K. Cooperatives and their primary care centres; organisation and impact. *Health Technology Assessment* (1999) 3 (7).
61 Hallam L. The integration of services in Salisbury C, Dale J and Hallam L (Eds) *24-hour Primary Care* (1999) Radcliffe Medical Press, Abingdon.
62 DoH. *Developing Emergency Services in the Community: Vol 1 Emerging Conclusions* (1997) NHS Executive.
63 DoH. *Developing Emergency Services in the Community: The Final Report* (1997) NHS Executive.
64 Donaldson L. Telephone access to health care: the role of NHS Direct. *J Roy Coll Phys Lond* (2000) 34: 33–5.
65 Jones R. Self care. *BMJ* (2000) 320: 596.
66 Munro J, Nicholl J, O'Cathain A and Knowles E. Impact of NHS Direct on demand for immediate care: observational study. *BMJ* (2000) 321: 150–3.
67 Salisbury C and Boerma W. Balancing demand and supply in out-of-hours care in Salisbury C, Dale J and Hallam L (Eds) *24-hour Primary Care* (1999) Radcliffe Medical Press, Abingdon.
68 Salisbury C. Out-of-hours care; ensuring accessible high quality care for all groups of patients. *Br J Gen Pract* (2000) 50: 443–4.
69 Page S. Introduction in Coffey T and Mythen M (Eds) *NHS Frontline – Visions for 2010* (2000) The New Health Network, London.

70 Lambert M. The A&E Modernisation Programme in Coffey T and Mythen M (Eds) *NHS Frontline – Visions for 2010* (2000) The New Health Network, London.
71 Department of Health. *Reforming Emergency Care* (2001) Department of Health, London.
72 Sakr M and Wardrope J. Casualty, accident and emergency, or emergency medicine, the evolution. *J Accid Emerg Med* (2000) 17: 314–19.

11 Politics and the Future

1 Department of Health. *The Patient's Charter* (1991) Department of Health, London.
2 Department of Health. *Reforming Emergency Care* (2001). Department of Health, London.
3 Anonymous. *See and Treat* (2002). NHS Modernisation Agency, London.
4 Anonymous. *Findings and Recommendations Form the Hospital at Night Project* (2004). NHS Modernisation Agency, London.
5 Binchy J. Accident and emergency medicine – the next 25 years. *J Accid Emerg Med* (1999) 16: 48–54.

Index

Abson Mr Edward 2, 33, 37, 44, 51, 109, 156, 157
Academy of Medical Royal Colleges 108
Accident and Emergency
  Academic development 100–103
  Name of specialty 1–3, 29–30, 36, 108–110, 139
Accident and Emergency consultants
  24 hour presence 77–79, 87, 88–91, 155
  Academic development 100–103
  Difficulties after appointment 68, 70–71
  First consultants – pilot scheme 47–49, 51–54
  Function of 30, 45–46, 78–79, 91–93
  Higher qualifications for 52, 53, 54, 55, 57, 59–60, 98–99
  Inappropriate appointments 54–56, 59, 81, 83
  Moratorium on appointments of 59, 60, 63, 112
  Numbers 74–86
  Specialisation 94–99
  Two consultant departments 61
  Workload of 82, 93–94
Accident services 6–8, 24, 27–28, 41
Accident Services Review Committee 28, 40
Achieving a Balance 83, 116
Acute medicine 98–99
Adams Mr William 66
Advanced Life Support courses 82
Allen Mr Michael 97
Ambulance Service 96–97, 153
Anderson Mr Ian 105, 158
Appointments advisory committee (AAC) 60, 63, 85, 103, 107
Associate specialist 111–113
Audit 101

Baderman Dr Howard 24, 50, 52, 62, 65, 103, 104, 105, 158
Baskett Dr Peter 105, 158
Beck Dr E 105
Bodiwala Mr Gautam 105, 106, 157, 158, 159
British Accident and Emergency Trainees Association 67
British Association for (Accident and) Emergency Medicine 87, 91, 110
British Medical Association (BMA) 32, 33, 37, 46
  Representative Body 47
British Orthopaedic Association 7, 27, 36, 38, 41, 43, 51, 54, 57–59, 87, 130, 133
British Paediatric Association 65
Browse Prof Norman 106, 158
Bruce Report 48
Bryce Miss Gillian 97

Calman Report 72, 85, 91, 107
Caro Mr David 5, 37, 42, 44, 51, 52, 156, 157
Casualty departments
  Consultant involvement 21–26
  Types of 4
Casualty Surgeons Association 36, 37, 52, 64, 65, 69, 74, 77, 80, 104, 108, 115, 143
Certificate of Completion of Specialist Training (CCST) 72, 98, 108
Chamberlain Prof Douglas 101
Clarkson Mr Patrick 3, 4, 21–22, 101, 123, 135
Clinical assistant 113–117
  Nine session clinical assistants 115–117
Coates Mr Timothy 97
Collins Mr John 51, 96, 156, 157
Committee on Fractures 7
Cooke Mr Mathew 88, 132

Court Report 142, 143
Cutting Mr Christopher 54, 157

da Costa Mr G 32, 33, 42
Dale Dr (later Prof) Jeremy 144
Dallos Dr Vera 77
Davies Mr Gareth 97
Dean Mr A 106
Developing Emergency Services in the Community 148

East London A&E Consortium 90
Edbrooke Dr David 97
Ellis Mr Maurice 22, 30, 31, 37, 40, 48, 50, 101, 156, 157
Emergency Care Practitioners 153
Emergency Medicine 109–110
Emergency Medicine Research Society 103, 107
European Working Time Directive 86, 118, 153

Faculty of Accident and Emergency Medicine 91
  Examination Committee 106
  Formation of 103–108
Fergusson Mr David 67
FFAEM exam 72
  Intercollegiate Board 105, 106
Fisher Dr Judith 146
Flowers Mr Michael 54, 56, 61
Flying squad 96
Forum for Associate Specialists and Staff Grades in Emergency Medicine 120
FRCS exam 40, 53–54, 68, 104, 105, 128
FRCS (Ed) exam in A&E 69, 70, 73
Future Strategies Group 105

Garden Mr 22, 45, 123, 138
General Practitioner Cooperatives 146–149
General practitioners working in A&E 10, 39, 42–43, 113–115, 123, 144–145
Gibson Mr Myles 69
Greet and treat 153

Hall Mr Malcolm 67, 104
Hardy Dr Richard 142
Head injuries 95
  Royal College of Surgeons Report on 95
Helicopter emergency service 97
Henry Prof John 102
Hide Mr T 106
Hindle Mr J 37, 156, 157
Hospitals
  Aberdeen 17, 51, 66
  Ancoats (Manchester) 7
  Bath 96
  Birmingham Accident 4, 8, 24, 52, 54, 56
  Birmingham General 38–39
  Bradford 17, 18, 25
  Bristol Royal Infirmary 80
  Canterbury 51
  Cheltenham 91
  City (Birmingham) 91
  Derby 51, 64, 96
  Devonport (Plymouth) 5
  Dundee 123
  Edinburgh Royal 5, 69, 96, 102
  Epsom 91
  Freedom Fields (Plymouth) 5
  Glasgow Royal Infirmary 40–41, 65, 66
  Gloucester 91
  Greenbank (Plymouth) 5
  Guys 3, 4, 11, 16, 17, 135
  Hallamshire (Sheffield) 124
  Harrogate District 120
  Hartlepool 126
  Hereford 18, 142
  Homerton (London) 90
  Hope (Salford) 67, 102, 104
  Kings College 17, 135, 144
  Leeds General Infirmary 22, 50, 54, 61, 64, 67
  Leicester 97
  London 4
  Luton 25
  Manchester Royal Infirmary 103
  Middlesex 4, 24
  Newcastle 124
  Newham General 90

Hospitals – *continued*
  North Devon District
    (Barnstaple) 125
  North Staffordshire Infirmary
    (Stoke) 18, 58, 80, 90, 97
  Nottingham 58
  Oldchurch 130
  Plymouth 5, 52
  Portsmouth Eye and Ear 5
  Preston 17, 22, 51, 67, 104, 123
  Queen Alexandra's (Portsmouth)
    5, 18
  Queen Mary's (Sidcup) 17
  Radcliffe Infirmary (Oxford) 4, 7,
    16, 24, 80
  Royal (Wolverhampton) 113, 146
  Royal London 90, 97, 146
  Royal Northern (London) 4
  Royal Victoria (Belfast) 18, 54, 62,
    67, 80
  St Bartholomew's 3, 5, 51
  St Mary's (London) 5, 23–24, 102,
    103, 145
  St Mary's (Portsmouth) 5
  Sheffield Childrens 95
  South Tyneside 120
  Sunderland 22
  Taunton 54
  University College 24, 50, 66
  Walsall 48, 51
  West Cornwall (Penzance) 124
  Western Infirmary, Glasgow 67, 97
  Weston super Mare 5
  Whipps Cross 77
  Worthing 120
Hospital at night 86
Hospital practitioner 114
House officer 10, 123–125

Illingworth Dr Cynthia 95
Inappropriate attenders 140,
  149–150, 153
Intensive care 97–98
Intercollegiate Board 105, 106
Irving Prof Miles 67, 101, 102

Jackson Dr R 95
James Prof JIP 55, 59, 65
Jennett Prof Bryan 101

Joint Committee on Higher Medical
  Training 65, 103 107
Joint Committee on Higher Surgical
  Training 65, 103, 107, 108
Joint Committee on Higher Training
  in Accident and Emergency 65,
  107–108
Joint Committee on Higher Training
  in Anaesthetics 65
Joint Consultants Committee (JCC)
  36, 48, 59, 62, 64
Jones Robert 7
Junior doctors (*see also* house officer,
  senior house officer)
  Hours 39
  Shortage of 11–12, 15, 38–39,
    125–131
Junior Hospital Medical Officer 9

Kirby Maj Gen Norman 105, 106,
  157, 158, 159

Laing Mr Gordon 67, 104
Lamont Mr D 13, 30, 31
Langston Mr H 33
Lewin Mr Walpole 33, 34, 59
Lewin Report 53, 59–60, 62, 67,
  69–70, 75, 78, 81, 104, 114, 130
Little Dr Keith 69, 96, 101, 102, 103,
  105, 106, 157, 159
Little Prof Roderick 102, 103, 107
London Mr Peter 54, 55, 140
Lord Mr Stuart 66, 157
Lowden Mr TG 4, 6, 8, 10, 22, 88,
  101, 132

Mackway-Jones Prof Kevin 103
Major Trauma Outcome Study
  (MTOS) 102
Manpower Committees 63, 67
Marsden Mr Andrew 67, 96
Matheson Mr Alasdair 66
Medical Act 1978 127
Medical Assistant 111–113
Medical patients in A&E 46–47,
  98–99, 154–155
Medical students 11, 38, 100–101
Memorandum on Accident Services
  (1959) 27–28, 36

Merlin Mr Michael 48, 51
MFAEM exam 73
Miles Mr Stephen 68, 105, 106, 157, 158
Mills Report 56, 62, 67, 70, 75, 114, 143
Minor Injuries Unit 88, 146, 149, 150, 152, 153
Moore Miss Fiona 96
Morgan Mr William 57, 67
MRCP 54, 55, 66, 68, 98, 104, 130
MRCS (Ed) in A&R 73
MRC Trauma Unit 101–102
Murray Mr Alec 40–41, 48, 51, 109

Name of specialty 1–3, 29–30, 36, 108–110, 139
National Audit Office 82
National Insurance Act 1911 135
Nelson Dr Peter 62
New Deal 91, 127, 130
NHS Direct 148–149, 150
NHS Modernisation Agency 153
NHS Plan 121
Nimmo Dr J 106
Non consultant career grade doctors 111–121
Non standard grades of medical staff 120–121
Nuffield Report 4, 5, 12, 15, 22, 28, 40, 88, 137–138
Nurse practitioners 130, 145, 154
Nurses in A&E 125–126

O'Connor Dr Peter 105
O'Connor Mr B 42
O'Higgins Prof N 105
Orthopaedic objections to A&E as a specialty 34–36, 43–44, 51, 52, 54, 57–59

Paediatric A&E 95–96
  Training in 96
Page Prof Graham 102
Pantridge Prof 101
Pascal Mr K 32
Patient's Charter 151
Patterson Dr 12

Platt Report (Sir Harry Platt Report) 1, 28–31, 32, 33, 44, 46, 88, 98, 108, 111, 123, 138, 139
Platt Report (Sir Robert Platt Report) 1, 25–26, 30, 32, 34, 111, 116, 123
Platt Sir Harry 7
Prehospital Care 96–97
Primary Care Emergency Centres 146–148
Primary care in A&E 3, 46, 114, 131, 135–150
  Consultants in 95, 145–146
  Patients' reasons for attending A&E 136–137
Proctor Mr David 51, 156
Pyke Dr David 106

Redmond Prof Anthony 102
Reforming Emergency Care 118, 150, 152
Registrar 69–70
  Work of 71–72
Research 101–103
Resuscitation room 42
Richardson Sir John 48
Robertson Dr Colin 102, 105
Robertson-Steele Dr Iain 146
Rosenheim Sir Max 46
Royal College of General Practitioners 65, 104, 138
Royal College of Physicians (London) 46, 64, 65, 99, 103, 105
Royal College of Physicians and Surgeons (Glasgow) 105
Royal College of Surgeons (Edinburgh) 69, 104, 105
Royal College of Surgeons (England) 48, 69, 81, 82, 95, 103, 104, 105, 128
  Report on patients with major injuries (1988) 82, 87, 90
Royal Commission on the Health Service 143, 144
Rutherford Mr William 30, 54, 62, 67, 69, 79, 103, 157
Ryan Prof James 102

Scott Mr J 7, 142
See and treat 153
Senior Casualty Officer 12–15, 31–32
Senior Casualty Officers' subcommittee of the CCSC 32–37, 45
Senior Hospital Medical Officer 9
Senior House Officer 125–134
  Origins of 131–132
  Workload of 132
Senior registrar
  Adequacy of training 68, 70
  Number of posts 66, 67, 75, 80, 83
  Qualifications of 66, 68, 70
  Training 66
  Travelling Club 67
Short Report 61, 76–77, 78, 79, 91, 115
Simpson Mr 66
Skinner Mr David 88, 158, 159
Snook Dr Roger 96, 101
Specialist Advisory Committee 55, 61, 65, 83–85, 96, 107, 108
Specialist Register 72, 85, 113
Specialist Training Authority 107–108
Specialist Workforce Advisory Group 85
Spens Committee 9
Sports medicine 97

Staff grade doctors 117–120, 130
Staffing of departments 8–21
Stewart Mr Ian 52, 56, 157
Stoddard Dr J 106
Stoner Prof B 102
Swann Mr Ian 66

Targets in A&E 152–153
Thrombolysis 98
Thurston Dr John 105, 107, 157, 158, 159
Todd Report 42, 128
Tomlinson Report 144
Touquet Prof Robin 103
Trauma Audit Research Network (TARN) 102
Triage 5, 143–144, 151
Trust grade doctors 120–121
Tullett Dr William 97

Vaughan Dr Gerard 52

Wardrope Mr Jim 97
Watts Mr J 40
Williams Dr David 24, 104–106, 157, 158
Wilson Mr David 40, 50, 54, 55, 56, 61, 65, 67, 101, 105, 157, 159
Wyatt Mr Jonathan 72

Yates Mr (later Prof) David 67, 92, 102, 103, 105, 106, 158, 159

DATE DUE

GAYLORD PRINTED IN U.S.A.